Biofertilizers and Biofungicides

Biofertilizers and Biofungicides

Dr. Ghazala Yaqub

Delve Publishing

www.delvepublishing.com

Biofertilizers and Biofungicides

Dr. Ghazala Yaqub

Delve Publishing

2010 Winston Park Drive,

2nd Floor

Oakville, ON L6H 5R7

Canada

www.delvepublishing.com

Tel: 001-289-291-7705
001-905-616-2116
Fax: 001-289-291-7601

Email: orders@arclereducation.com

ISBN: 978-1-77361-253-9 (Hardcover)

ABOUT THE AUTHOR

Dr. Ghazala Yaqub

Dr. Ghazala Yaqub is an Assistant Professor of Environmental Sciences and a member of Academic Council at Kinnaird College for Women, Lahore, a well reputed and prestigious institute in Pakistan. She graduated with a Ph.D. degree in Chemistry in 2011. She did her Post-Doctorate research at the University of Wollongong Australia. She is an author of various research articles and books. Dr. Yaqub was awarded Australian Excellence Award in recognition of her research outputs and community services in 2016. She is recently appointed as an Australian Award Ambassador for Pakistan.

TABLE OF CONTENTS

PREFACE

Chemical fertilizers are widely used with the purpose to enhance output in various product varieties in agriculture. As, chemical fertilizers are the biggest reason of posing pollution to the water bodies, ground water, and also bioaccumulates in crops hence badly destroying ecological cycles. Therefore, modern scientists are focusing to switch completely from chemical farming to organic farming in order to encourage sustainable agriculture. In organic farming crops are not polluted as they rose through use of manures, biofertilizers and biopesticides which not only provides optimum nutrients to plants but also keep pests and pathogens in control. Biofertilizers are actually "microorganisms which bring roughly nutrient enrichment of soil through enhancing the availability of nutrients to plant crops". These micro-organisms who exhibit the ability to act as bio-fertilizers are the bacteria, cyanobacteria and *mycorrhizal* fungi. Biofertilizers provides considerable potent benefits to agriculture as well as public health programmes. Both developing and developed countries have a vast potential for biofertilizers. However, their adoption by farmers especially in developing countries needs education for maximizing benefits. The stress on organic farming as well as on residue free materials would certainly merit enhanced adoption of biofungicides by the farmers. The use of fungicides reduces crop productivity loss because the chemically synthesized pesticides are no doubt quick in their effect but on the other hand they cause so many environmental and health just as revival of new pests, fungal resistance, loss of soil fertility and persistence of toxic substances when they are consumed by animals and humans. All these problems call for an eco-friendly and sustainable approach in order to decrease the usage and dependence of chemically synthesized fertilizers and pesticides. Similarly, there are so many natural fungicides that can be best alternatives to control fungi and enhancing plant life. Thus, the purpose of this book is to provide that knowledge about recent advancements in the emerging fields of Biofertilizers and Biofungicides as both of them are considered to be ecofriendly as well as sustainable substitutes to chemical fertilizers and fungicides respectively.

Chapter 1- 10 discusses all aspects of biofertilizers with special focus on recent advances in this technology while Chapter 11- 14 are about Biofungicides, their applications and recent advances in this field of organic farming.

Author

1 CHAPTER

INTRODUCTION TO BIOFERTILIZERS

INTRODUCTION

A substance which increases the availability of nutrients to the plant when it is applied to the soil is known as Bio-fertilizer. They increase the nutrients availability to the plant through numerous processes i.e. solubilizing phosphorous, nitrogen fixation and synthesis of growth promoting substances. Soil's natural nutrients are restored by bio fertilizers due to which soil becomes more fertile and sustainability of soil increased. Ecofriendly organic agro input is provided by bio fertilizers .Common bio fertilizers are Azobector, Rhizobium, Blue green algae. Production of bio fertilizers is increased by the government and research organization due to the demand by the farmers. In case of bio fertilizers chances of water pollution is lesser. Bio-fertilizers are cost effective relative to the chemical fertilizers.

The biota living in soil is of great value to the farmers and also the ones interested in farming and gardening. The soil organisms make it possible for plants to absorb and utilize the nutrients such as Nitrogen, Phosphorous and Sulphur. They regulate the decomposition mechanism in the soil. Without the beneficial microbes in soil the agriculture is impossible. They benefit in increasing the crop productivity and eliminate the excessive usage of fertilizers on the land.

Bio-fertilizers are the solution to stop the destruction of soil structure, water quality and flora. They are organic and all natural fertilizers. They fall under sustainable organic farming. They are the best alternatives for replacing the conventional chemical based fertilizers.

Biofertilizers can be defined as

"A substance which includes or involves the living organisms or microorganisms and is helpful in promoting the growth of the plant's root system and germination of seeds"

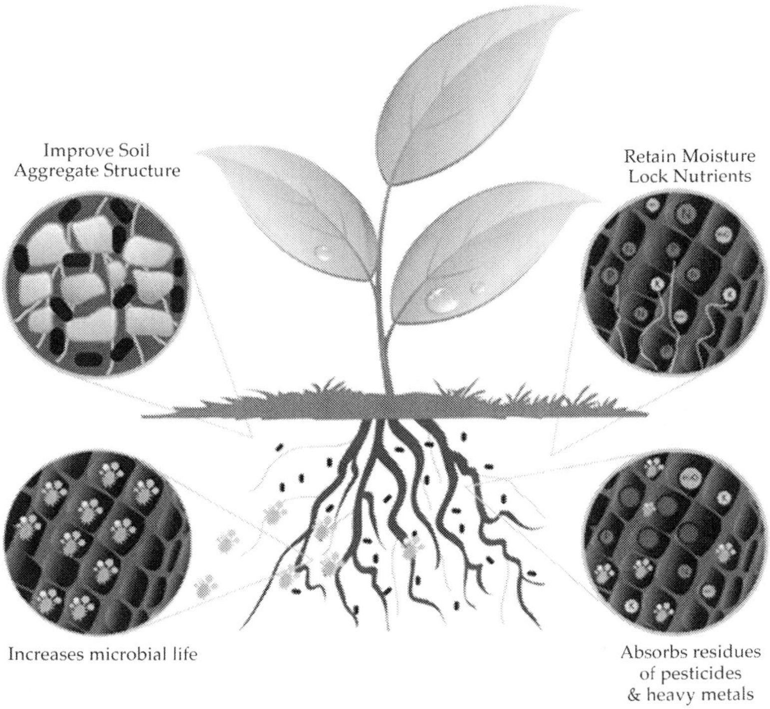

Figure: Biofertilizer-benefits (source: https://www.linkedin.com/pulse/what-biofertilizer-dr-tohid-nooralvandi)

Constraints in bio fertilizer technology

Most of the bio fertilizer have common characteristic of the unpredictability of their performance. In order to gain benefit from the bio fertilizer there is need to improve their consistency in agriculture. Though bio fertilizers are eco friendly and low cost technology but some constraints still exist that limit their application and implementation of technology. These constraints may be financial, economical, human resources and quality, environmental, infrastructural, production and technological which effect the marketing production and use of technology through one way or other.

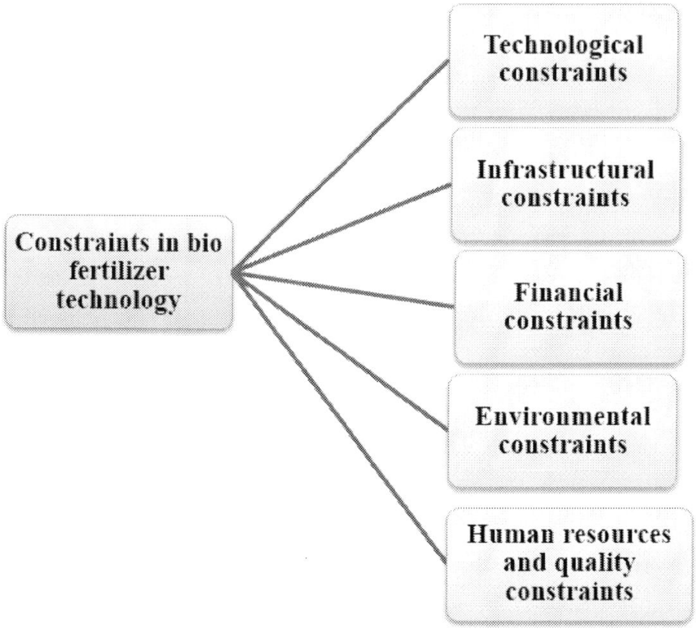

Figure: Constrains in bio fertilizer technology

Technological constraints

Constrains in technology include usage of least effective and improper strains for formulation as well as lack of technically qualified staff in the production units, use of poor quality material or taking material from different producers without knowing their quality, un availability of good quality material and production of bad quality inoculants because of the non-understanding of microbial technologies.

Infrastructural constraints

Lack of availability of facility for production or lack of space for production, storage and laboratory are included in infrastructural constraints. Sometimes space is available but lack of equipment and power supply become constraints in production. For better production availability of facility for cold storage of inoculants packets are also required.

Financial constraints

Lack of funds or problem in getting loan from banks, less return by sale of products and higher capital cost of technology in smaller production units are the financial constraints.

Environmental constraints

These constrains include seasonal demand of bio fertilizers, concurrent cropping operations, Very short life span of sowing/planting in a particular locality, soil characters like salinity, acidity, drought, water logging, etc. are environmental constraints.

Human resources and quality constraints

Another greatest constrain in bio fertilizer technology are human resources and quality constrains as there is unavailability of trained personals to use and adopt this techniques in field as well as no appropriate training on the production procedures and protocols is available. In many cases product quality is totally ignored as quality specifications and quick quality control methods are not available to larger extent. These factors are intern responsible for poor quality products. This problem of adaptation of technology of using biofertilizers also arises at farmers end when they use different methods of inoculation. It's also one of the greatest quality as well as human resource constrain that people are not completely aware of this technology and potential benefits of using it as they found no visible difference in the crop growth directly as that of inorganic fertilizers

Use/ Application Biofertilizers

Bio-fertilizers are not among the ordinary fertilizers. They are those specific of fertilizers which are composed of microorganisms. The applications of these biofertilizers to different parts of the plant like seed, soil, plant surface and even to the interior parts of the plant i.e., rhizosphere accelerate and promotes the growth as they enhance supply and availability of most of the primary nutrients.

Unlike conventional fertilizers who act by directly increasing the soil's fertility by addition of different nutrients from outside, bio-fertilizers exhibit the ability to add necessary nutrients via following natural processes i.e., (i) fixing atmospheric nitrogen, (ii) solubilizing Phosphorus and last but not leasy by (iii) stimulating plant growth by synthesizing growth promoting substances.

In past few decades upon the introduction of green revolution technologies, the modern agricultural practices are getting much dependent upon the secure supply of different synthetic inputs (mostly fertilizers). As most of these fertilizers/ synthetic inputs are the products of fossil fuels i.e., coal + petroleum etc. they exhibit full potential to pose adverse effects which are fully attributed to their excessive and imbalanced useage. This senerio provides the basis of identifying and using some harmless inputs like biofertilizers. As the Crop which are cultivated by using these natural products i.e., biofertilizers not only help is protecting the soil and health quality but also the quality of produced crop.

Bio-fertilizer application methods

There are three ways of using Bio-fertilizers

1. Seedling Root dipping
2. Seed treatment
3. Soil application

1. Seedling Root dipping

This application procedure is adopted for transplanted crops like paddy, tobacco, onion, cauliflower, tomato, chilly etc. The following methodology is used for this treatment:

A suspension is prepared by mixing 1 kilogram of bio-fertilizer culture in 10-15 liters of water. Seedlings required for one-acre land are collected and small bundles of those seedlings are formed. The root portion of these seedlings is dipped in the suspension prepared earlier for 15-30 minutes and then transplanted immediately. Generally, the ratio of inoculant and water is approximately 1:10 which means one kilogram of bio-fertilizer in 1- liter of water.

For application of Azospirillum/ /PSM (**Phosphate solubilizing microorganisms)** on paddy/ vegetable crops this method is used. The required quantity of Azospirillum/ /PSM has to be mixed with 5-10 litres of water at one corner of the field and the roots of seedlings has to be dipped for a minimum of half-an-hour before transplantation.

Figure: Seedling Root dipping (source: http://www.ecotreecare.co.uk/mycorrhizal-inoculation.htm)

These bio-fertilizers can also be used for ornamental plants such as jasmine, roses, dahlias etc. Regarding this method of application, it is usually done by root dip method or by the cutting method. For this, mix 200 grams of bio-fertilizer in 2 liters of water, which is enough to treat 200 to 300 plants. In a similar fashion 200 grams of bio-fertilizer in 2liters of water is sufficient to treat 200 to 300 plants by cutting method.

2. Seed treatment

Seed Treatment is a most common method adopted for all types of inoculants. The seed treatment is effective and economic. For small quantity of seeds (up to 5 kilograms) the coating can be done in a plastic bag. In order to fulfill this purpose a plastic bag may be used which can be of 21 x 10cm in size. Then for seen treatment the selected bag is filled with 2-3kgs of targeted seeds and then the bag is ultimately sealed in such a way to trap as much as possible air in it. Then this bag containing seeds is compressed and squeezed for about 2-3 minutes to ensure uniform wetting of all seeds. This process of opening the bag to trap more air is repeated again and again until it is ensured that each seeds acquired a uniform layer of that culture coating. Once all seeds have uniform layer the plastic bag is opened and seeds are allowed to dry below some shade for half hour. For greater quantities of seeds the above described coating can be done in some bucket and inoculant to be used can be directly mixed with hands. The similar method of treating Seeds can also be adopted with Rhizobium, Azotobacter, Azospirillum etc.

Figure: seed treatment using biofertilizer (source: http://shramajeewiki.com/English/Organic-Farming-part-1.html retrieved on 10-05-2017)

The process of seed treatment may be done by using two or more bacteria. There are ultimately zero side/ antagonistic effects. The utmost important point to be kept in mind while seed treatment is that all targeted seeds should be primarily coated by Rhizobium, Azotobacter or with Azospirillum. Once each targeted seed got a layer of above described bacteria then in next step PSM inoculant should be coated as its outer layer. This method will make available each of the bacteria which is required to have best results. Maximum number of bacteria on individual seed can never be obtained by treatment of seed with any two above said bacteria will not provide.

The explained method of seed treatment is best to be used for pulses/ oil seeds and fodder crop with few modifications. One pack of the inoculant is stirred with about 200 milliliter rice kanji in order make slurry. The seeds which are required for a single acre are stirred well in the slurry in order to have uniform and equal coating of the inoculant over seeds followed by its drying under shade for half hour. Then these shade dried seeds are to be sown within a day. One packet of the inoculant is enough to treat about 10 kg of targeted seeds.

3. Soil Treatment

Method of soil treatment greatly varies for crop to crop depending on the duration of crops. Generally, for short duration crops i.e., those less than 6 months, only 10 to 15 packs are mixed with about 40- 60 Kgs of the decomposed animal manure or it may also be mixed well with about 40-60 kilograms of soil for about one acre land.

This obtained mixture of bio-fertilizer + cattle manure/soil is scattered with water is then it is added in targeted soil when there is time of sowing or irrigation in standing crops. All of the bio-fertilizers along with compost fertilizers are mixed well together and are kept overnight. Next day this mixture is spreaded on targeted soil where seeds are required to be sown. While for long duration crops (perennial crops) about 20 - 30 packs of bio-fertilizer are mixed with 80 - 120 kilograms of manure or soil per acre.

Figure: Soil Treatment (Source http://rsfpdkalyani.com/biofertilizermain.php?id=66%20&%20category=Bio-fertilizer)

Role of Bio Fertilizer in Agriculture

Due to the natural disasters and heavy causalities i.e. drought, high temperature, cold damage, agricultural land is badly affected and ultimately it affects the security and lives of residents. Therefore establishing environment friendly mechanisms is very important. There must be the effective ways to achieve this goal. In past years pesticides and chemical fertilizers were used to increase the fertility of soil and maintain high crop yield but chemical fertilizers increased ground water contamination, soil acidification and weakened the roots of plants and made them more susceptible for different diseases. On other hand there are bio fertilizers which strengthen the roots of plants, provide them essential nutrients and increased the productivity of the soil. Thus, they provide the better conditions to maintain the soil fertility than chemical fertilizers. The reason of choosing bio fertilizers over chemical fertilizers by researchers and farmers is Environment Friendly behavior.

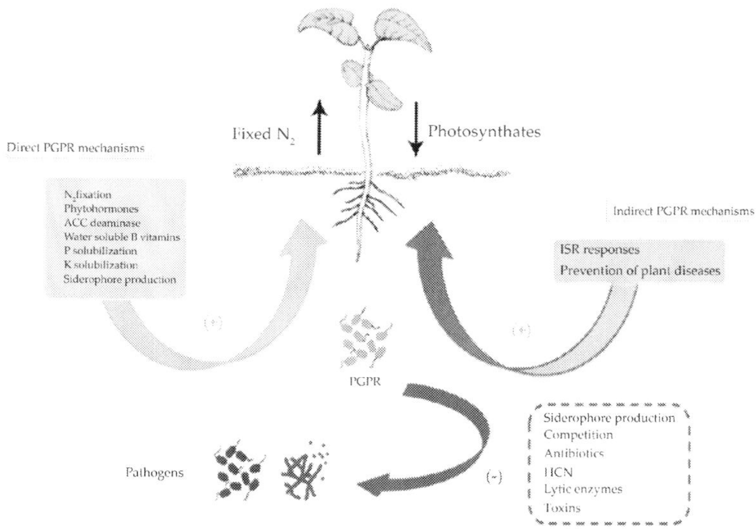

Figure: Role of bio fertilizers (source: Paula García-Fraile, Esther Menéndez, and Raúl Rivas, Role of bacterial bio-fertilizers in agriculture and forestry, AIMS Bioengineering, 2 (3): 183-205, 2015)

Yield Promoters

Biofertilizers are one of the best yield promoters. Uptake of phosphorous by the roots of plants is increased by using bio fertilizers. There are several species of plant bacteria which grow in the rhizosphere of plants, enhanced the plant growth by the mechanism of plethora. These bacteria known as plant growth promoting rhizobacteria. These bacteria increased the availability of nutrients in rhizosphere, fixing of nitrogen, root growth and provide the symbioses relationship thus enhanced the yield of plants.

Future of Bio fertilizers

Farmers are adopting the bio fertilizers in their fields to avail the benefits from it. Adoption of bio fertilizers depends on the knowledge. The researchers are working on the adoption level, knowledge level, problems and failure of bio fertilizers. Recommended fields for the bio fertilizers are paddy and pulses. A Survey was conducted and it was concluded that half of the respondents were using the bio fertilizers and they were having the knowledge about it. Some of the farmers were using one kind of bio fertilizers while some of them were using more than one type of bio fertilizers .Reasons for the non-adoption were lack of conviction, lack of time, lack of will to take the risk and lack of difficulty to understand the terminology. If farmers get the proper information about the bio fertilizers their future will be dazzling with opportunities.

Organoxeolitic Biofertilizer

To produce the soil substrate animal waste is used and zeolitic tuff contains it. Accrding to the study it was concluded that the leachate samples of organic soil mixtures which are having only animal manure contains less concentration of nitrate than the samples which are having zeolitic tuff. In Clinoptilolite there is an ion exchange property due to which it is selective towards the ammonium and it appears that clinopilolite component in ion exchanging ammonia as it is liberated from decomposing animal waste due to which ammonium oxidizing microorganisms are stimulated and enhanced Nitrification. Due to this function of organoxeolitic bio fertilizer, plant can be grown in acid mine sites which is normally barren land and the mechanical analysis of roots and stem proved the uptake of nutrients in that land due to the usage of organoxeolitic microorganisms.

Managing plant diseases with bio fertilizers

Several PGPR bio inoculants are currently used commercially to manage plant diseases. They are called different names and operate through a range of mechanism.

- Bioprotectants, through suppression of plant disease
- Biofertilizers, through improved nutrient acquisition
- Biostimulants, through phytohormone production

Bacteria which are included as bio inoculants include range of bacteria belonging to the genera Bacillus, Paenibacillus Streptomyces, Pseudomonas, Burkholderia and Agrobacterium. They are use at commercial level in order to suppress diseases of plants. The plant diseases are suppressed via induction of systemic resistance and production of siderophores or antibiotics as well.

Biofertilizers are available to increase crop uptake of nitrogen from nitrogen-fixing bacteria like *Azospirillium*, and also to uptake iron from siderophore-producing bacteria as for example *Pseudomonas*. Species of both Pseudomonas and Bacillus have the ability to produce (either not well characterized) phytohormones or the growth regulators that have the ability to accelerate extensive root growth by increasing the absorptive surface of the targeted plant roots. These PGPR are referred to as biostimulants and the phytohormones that they produce include indole-acetic acid, cytokinins, gibberellins and inhibitors of ethylene production. Existing modes of delivering inoculants include peat, granular, liquid as well as wettable powder formulations. A main determinant of growth promotion is greater magnitude of their ability in order to colonize rhizosphere.

Environmental Limitations for Application of Bio-fertilizer

There are the following environmental limitations for the application of biofertilzers:

- Lack of awareness among the farmers.
- Lack of quality assurance.
- Social factors and climatic factors.

Formulations of bio-fertilizers-approaches and advances

Bio-fertilizers are an emerging source for sustainable agriculture. It nurtures the soil with useful microorganisms. Microorganisms break down the organic matter and re-stock nutrients into the soil.

Bio-formulation is a carrier in which microorganisms, latent or live (bio-inoculants) are delivered to the targeted plant. The formulation helps easy handling, effectiveness and storage of microorganisms. The bio-inoculants technology is successful because of two factors, microbial strain and the formulation of inoculants. The technology of formulation is independent of strain of microbes. This is because they share physiological properties. In this case the technology of a strain can be used for another strain with little modifications.

Bio-inoculants used are of many types. Some of them are listed below:

- N fixers
- P solubilizers
- P mobilizers
- Biocontrol agents
- PGPR

Before application the bio-inoculants must be put in a solid or liquid based carrier along with osmoprotectant, nutrients and sticking agents; this complete mixture is called a formulation. Formulation can vary according to the type of soil on which it is being used, type of plant, resource availability and method of application. Development in formulation is gaining momentum majorly towards product development.

A good formulation provides optimum conditions to elevate the persistence of organisms in soil and to obtain maximum benefit from their activity after they are applied to the soil. The inoculants must be easy to handle and cost effective. Each formulation has its own advantage and disadvantage/limitations.

Table: Advantages and limitations of carriers

Carrier	Advantage	Limits
Peat	Suitable for a wide range of microorganisms: bacteria, AMF , ectomycorrhiza Protective nutritive environment Moisture content can be adjusted to optimize growth and survival of bacteria during curing, storage and inoculation Strong buffering capacity	Not readily available Strong negative impact on environment and eecosystem Costly investment for extraction Toxic compounds released during drying and sterilization Highly variable in composition and quality depending on the origin Seed application, contact with other chemical compounds

Liquid	Easy to handle and apply Easy addition to additives to improve growth or survival of cells Composition easily defined and controlled High cell concentration-low application rates	Lack carrier protection: low viability during storage and on seeds Cool temperature for storage (4°C) Limited shelf life More sensitive to stressful conditions
Granules	Easy to store and handle and apply Less dusty than peat Application rate easily assessed Soil application: no direct contact with the other chemical compounds (no toxicity) Especially efficient under stressful environmental conditions	Bulky: high transport and storage costs Higher application rates Often non-sterile carriers
Encapsulated cells	Suitable for all types of cells Cells protected in a nutritive shell against mechanical and environmental stresses and against predators Slow and controlled release of the microorganisms when the shell is degraded Wide variety of polymers Storage at RTP (dried capsule)	High production cost More handling work at the industry level Specific equipment required Physiological, morphological and metabolic changes occurring in the shell Several applications needed if strains cannot establish in soil No commercial product available.

Inoculants

Inoculants are commercially prepared rhizobium bacteria. They are produced naturally in soil and are also applied artificially to the soil for nourishment.

Types of inoculants

There is a variety of inoculants that are listed below:

- Dry inoculants
- Granules
- Liquid inoculants
- Cell immobilization
- Other formulations

Dry inoculants

They are produced by organic or inert carriers. Most of the formulations are used using peat. The dry inoculants is sprinkled and mixed onto the seed. This is a fast way of seed inoculation.

Granules

Granular inoculants are made from peat prill or small silicate or calcite grains, wetted with an adhesive agent and mixed with powder inoculums. Then they are applied to soil.

Liquid inoculants

They are based on aqueous, mineral and organic oils, suspension and oil in water. They are easy to handle and apply. The inoculants are mixed with a liquid sticker. The slurry is then mixed with seeds. After coating they are spread out and are allowed to cool.

Cell immobilization

Immobilization caters different forms of cell attachment or entrapment into a matrix. These include flocculation, adsorption on surfaces, covalent bonding to carriers, cross-linking of cells, and encapsulation in a polymer gel. Encapsulation is proven as a most promising technique for constructing carriers of microorganisms, with substantial advantages over other types of carriers. Different kinds of polymers may be used for encapsulation for instance natural (polysaccharides, protein material) or synthetic (polyacrylamide, polyurethane) and homo-, hetero-, or co-polymers.

Other formulations

Sludge water is a good inoculation technique but the presence of heavy metals can raise a problem. Compost from cork industry is a good alteration source for peat.

Bio-fertilizers - A key player for sustainable agriculture

Conventional agriculture use fertilizers and pesticides to meet the growing food demand. Fertilizers are manipulated with known concentration of nitrogen, potassium and phosphorous and their use cause air and water pollution. In this regard efforts are made to introduce nutrient rich food while ensuring bio safety. Organic farming depends upon the microbes of the soil which include bacteria and fungi. Bio-fertilizers keep the soil rich with micro and macro nutrients. The processes that keep soil healthy are

Nitrogen fixation

It is a biological process in which atmospheric nitrogen is captured by bacteria and is converted to plant usable form. They enhance soil productivity and supplying sufficient nitrogen to the plants. They enhance mineral and water uptake by plants.

Phosphate and potassium solubalisation

It solubilizes inorganic phosphorous from insoluble compounds by the addition of solubilizes.

Mineralization

It decomposes or oxidizes chemical compounds in organic matter in plant accessible form.

They release and regulate substances by producing antibiotics and degrading organic matter. 60-90% of the fertilizers applied are lost while only 10-40% is taken up by plants. Bio-fertilizer is one of the viable technologies which are less explored by the farmers, due to its slow action, less-visible effects to the crops and less persistence in soil.

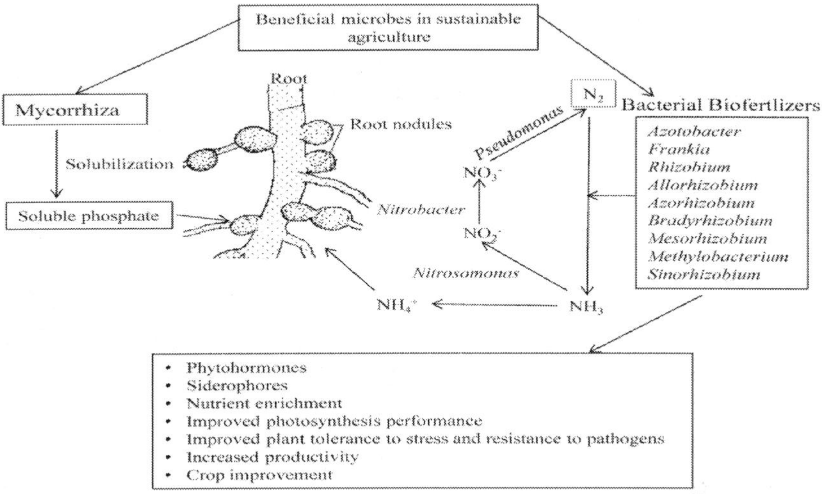

Figure: Use of microbes for sustainable crop production (Source: Deepak Bhardwaj, Mohammad Wahid Ansari, Ranjan Kumar Sahoo and Narendra Tuteja, Biofertilizers function as key player in sustainable agriculture by improving soil fertility, plant tolerance and crop productivity, Microbial Cell Factories, 13: 66, 2014)

Bio-fertilizers exploitation and nutrient profile of crops

Microorganisms assimilate phosphorous for their own requirement. Pseudomonas, bacillus, penicillium etc. are active in solublisation process. A phosphate solubilizing bacteria have the ability not only to solublize phosphorous but it is responsible for siderophore production. Similarly two fungus Aspergillus fumigates and A. Niger were isolated and were found converting cassava waste by fermentation to phosphate bio fertilizers. They can mobilize in different crops. Mycorrhizal mutualistic symbiosis with plant roots satisfies the plant nutrients demand which results in enhanced growth of plant and protects it against pathogens and environmental stress. It absorbs phosphate which is transferred to root cells.

Biofertilizers relevance and plant tolerance to environmental stress

Environmental stresses both biotic and abiotic are affecting the yield of the crop. Scientists have invented some modern tool to overcome this in which PGPRs is playing an important role as a bio protector. It has been seen in several instances that biofertilizers/ bio protectors are helping to save the crops from several types of environmental stresses. For example, *Trifolium Alexandanrium* has been to sustain salinity stress if inoculated with Rhizobium trifolii. Pseudomaonas aeruginosa have been seen to sustain both biotic and abiotic factors. Another biofertilzer *P.florosence* MSP 393 have been seen to produce salt stress induced protiens and osmolytes to counteract the negative effects of salt. It has been observed that AM fungus if inoculated to crop also provides protection to plant growth against salt stress. A fungus name *Piriformospora Indica* which is a root endophytic fungus has been seen to protect plant salt stresses. Similar effect was seen with numerous other fungus with or without PGPRs inoculated with it. From these examples it can be inferred that these biofertilizer have potential to take the place of chemical agents as not only increase the growth of plant but also produce certain metabolites that help the plant to defend themselves against pathogen and environmental stresses.

Mechanism of action of various biofertilizers

Mycorrhizal association of fungus with plant roots is an important model to understand the mechanism of action of several biofertilizers. While the gene sequencing of some EM fungi like T. melanosporum 14 and L biocolour 13, revealed a number of factors that help in the function and development of mycorrhizae in plant cell. There were 15 genes in L. biocolor that helped in the regulation of symbiosis known has putative hexose transporters. The genes were enable to encode inverters enzyme hence they were dependent on plant for glucose. However melanosporum had one invertase hence was able to take up sucrose of plant directly. The transporter gene up-regulation during symbiosis process showed that the useful compound like oligopeptides amino acids and polyamines moves from one organism to another through symbiotic interface. Nitrate and ammonium can be taken up free mycelium then these compounds reach hartingnet and mantle and are transferred to plant. The symbiotic interface is made with the help of Cystenic rich proteins which act as effectors and facilitators. Also some genes related to root morphogenesis and biosynthesis showed up-regulation during the development of mycorrhizae. Further, G versiforme and G, intraradice have phosphate transporter which help in absorption of phosphate and glutamine synthetize that supports the possibility of nitrogen metabolism respectively. Myc factors are some bioactive compounds found in mycorrhizae that are perceived by host for the initiation of common symbiosis pathway and signal transduction pathway. These pathways that braces the plant for rhizobium or mycorrhizae has some common points. They bring about anatomical and molecular change as the plant root hairs come in contact with fungus. Calcium is the hub of secondary messages that are sent via Ca 2+ which are spiking in the root hair's nuclear region.

Advanced technologies for using beneficial microorganisms inocula as modern biofertilizers

The practice of inoculating many plants with PGPM may be drawn back to the early eras 20th century, when for the first time a creation possessing *Rhizobium sp.* was well patented. Despite of the fact that *Mycorrhizal* fungi has been used as biofertilizers since few decades, it was late 1950s when they were confirmed to promote plant growth via P uptake. Since that time, exploratory efforts in these specific fields have progressively increased, which results, in most recent years, in collection of various strains showing quite a few significant features.

Formulation of Inocula

By using fermenters the preparation of selected yeasts as well as bacteria in pure cultures is a reasonably common and adopted practice. Thus, it is very easy to define a standardized manufacturing process once a particular strain for inoculum has been selected. While, in case of biofertilizers, unlike typical biopesticides, the cost of production is a very important constraint. As the cost of the fertilizer/biofertilizer should not exceed to the cost of conventional ones in order to make sure market sustainability. As many cheap organic matrixes (e.g., whey/ water sludges/ composts etc.) have been tested to act as growth media for PGPM, the bestpossible way to cut-short the production price is by adopting the use of agroindustrial residues which are significantly enriched with rock phosphate. Free or immobilized microorganisms which may produce organic acids are added to the matrix during composting or fermentation process in order to improve the solubilization of phosphate, which make it more available to plants.

Carriers

By weight as well as by volume, the carrier is one of the major part of the inoculant which helps in deliver a appropriate amount of PGPM in acceptable physiological condition. Materials of different origins i.e., (i) organic, (ii) inorganic, or (iii) synthesized constitutes the carrier. Availability and price are the main factors strongly affect the choice of a carrier. Two main types of carriers are provided below:

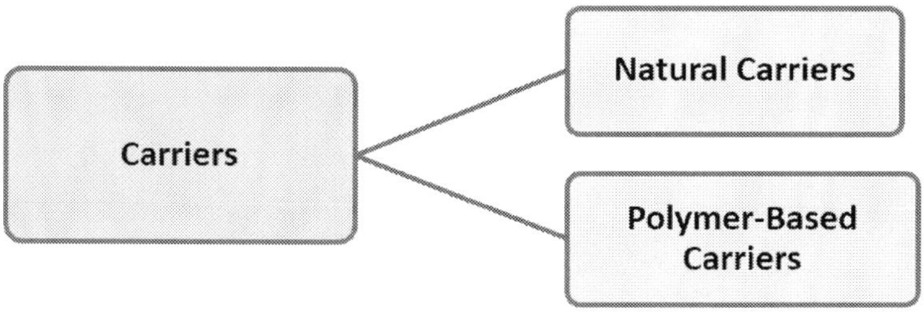

Figure: Catagories of Carriers.

Natural Carriers. Due to wider availability and strong background of field trials, Peat is mostly used as carrier for PGPR, predominantly for rhizobia inoculants. When added to peat, PGPR gain more ability to maintain metabolic activity and in majority of cases it may continue to multiply during the storage period resulting in increase in its population size but this can vary from strain to strain. While the other side of coin is the, a most important drawback of peat that is the variability in its quality as well as in its composition (acidity) this is attributed to its origin from different production sites, which have the potential to affect PGPM viability, in addition, peat embrace a huge load of microorganisms, which may reduce the shelf life of the inoculant. Similar disadvantages can also be attributed to other carriers which are made of plant waste materials that are generally not used for commercial preparations. Whereas, composts may also be considered as potential carriers, especially in case when the process of their production involves the usage of particular selected strain.

Polymer-Based Carriers. With the increasing interest in the application of bacterial preparations to act as plant protection products the researches which aimed at improving their stability and focused on increasing their shelf life have also been well promoted and documented.

Though, the similar approach could be productively applied to the products which contain PGPRs and AMF. Organic polymers have also been well documented among the new resources which may be utilized as carriers for PGPM. These are compounds (e.g., polysaccharides) who in the presence of ions or by changing their chemical conditions (e.g., a change in pH of the medium) form cross-links that create a complex structure. These polymers encapsulate, or "immobilize", the microorganisms in the matrix and then release them slowly via a degradation process. Polymer formulations not only offer a long shelf life (even at ambient temperature) while they also provide protection against possible environmental stresses and a reliable batch quality attributed to standardized production.

Biofertilizers - An ecofriendly source for sustainable modern agriculture

Biofertilizers is considered to be one of the ecofriendly substitutes to chemical pesticides. Biopersticides covers a broad array of (i) microbial pesticides, (iii) biochemicals derived from micro-organisms as well as (iii) other natural sources and processes including the genetic incorporation of DNA into agricultural commodities which bestow defense against pest damage. Biopesticides are categorized into three main classes. Biofertilizers provides considerable potent benefits to agriculture as well as public health programmes. The showed interest in biofertilizers by researchers and scientists is due to the disadvantages which are strongly associated with chemical pesticides/ fertilizers. Both developing and developed countries have a vast potential for biofertilizers. However, their adoption by farmers especially in developing countries needs education for maximizing benefits. The stress on organic farming as well as on residue free materials would certainly merit enhanced adoption of biofungicides by the farmers.

Increasing awareness about environmental conservation as well as potential health hazards associated with the use of agrochemicals and consumer's preference to adopt safe and secure food are the main factors which may lead to the increasing interest in alternate forms of agriculture of this world. Organic agriculture is no doubt the broad spectrum of production methods which are supportive to environment. The demand for the organic food is steadily increasing all over the world with an annual average growth rate of about 20–25%. Organic agriculture is obviously one of the fastest growing sectors of modern agricultural production.

Assessment of biofertilizers role in cultured rice

Biofertilizers, as renowned to have versatile applications, their effects on grain yield of cultured rice have also been studies. In order to see the benefits of biofertilizer's use, three different kinds of biofertilizers (which were based on Azospirillum, Trichoderma, or unidentified rhizobacteria) have been evaluated during four different cropping seasons between (2009-2011). Biofertilizers were used at four different fertilizer rates i.e., 100% of the recommended rate, 50% of the recommended rate, 25% of the recommended rate and ultimately no fertilizer as Control. Complete irrigated conditions in a typical lowland rice environment were adopted to perform the experiment. Considerable increase in yield was noticed due to biofertilizer use in all experimental seasons with the only exception of the 2008/09 DS. Though, the effect on rice grain yield varies between biofertilizers, seasons, and the fertilizer treatments. The seasonal yield which was noted to be enhanced across fertilizer treatments was between 5% to18% for the best biofertilizer (Azospirillum-based), but this was improved up to 24% in individual treatments. Absolute grain yield increases due to biofertilizer were usually below 0.5 $t{\cdot}ha^{-1}$, corresponding to an estimated additional N uptake of less than 7.5 kg N ha^{-1}. The tremendous effect of biofertilizer in enhancing yield did not considerably interact with inorganic fertilizer rate used and the finest effects on grain yield were acquired at low to medium fertilizer rates. However, positive effects of biofertilizers were even observed at grain yields up to 5 $t{\cdot}ha^{-1}$. Nevertheless, the trends in the obtained results are clearly indicating that biofertilizers are possibly be most supportive in rain fed environments with very limited inorganic fertilizer input. However, for use in such target environments, biofertilizers are needed to be assessed under conditions with much of abiotic stresses typical of such systems e.g., drought, low soil fertility and soil acidity etc.

Shelf life of biofertilizer an accord between formulation and genetics

Currently, a wide range of commercial biofertilizer formulations are accessible and many different strategies have been used to ensure the maximum viability of the microorganisms which are used in these formulations. These strategies are mostly comprised of:

- Optimization of biofertilizer formulation,
- Application of thermo-tolerant/drought-tolerant/genetically modified strains and,
- Application of the liquid biofertilizer.

To ensure their smooth application, a carrier material (already described in previous chapters) is used as a vehicle for the microorganisms which were to be used as biofertilizer. Furthermore, such materials have a strong potential to maintain the viability in terms of shelf life of the microorganisms before its release into the field. Other than this they also provide an appropriate microenvironment for rapid growth of the organisms when they are released. A carrier is a specific material i.e., (i) peat, (ii) vermiculite, (iii) lignite powder, (iv) clay, (v) talc, (vi) rice bran, (vii) seed, (viii) rock phosphate pellet, (ix) charcoal, soil, (x) paddy straw compost, (xi) wheat bran or a combination of any such materials. In general practice, for improved shelf-life of biofertilizer formulation, a carrier or a combination of such carrier materials are chosen based on the viability of the microorganisms mixed with these carriers. In the same way, pre-sterilization of the carrier material as well as its enrichment with the nutrients is the other promising strategy in order to improve shelf-life of biofertilizers by allowing the targeted microorganism to maintain and grow in some non-competitive microenvironment. Some of the supplementary nutrients and the cell protectants which are commonly used are (i) Sucrose, (ii) maltose, (iii) trehalose, (iv) molasses, (v) glucose and (vi) glycerol. These are used along with a carrier material in order to make sure maximum cell viability as well as extended shelf-life. Liquid biofertilizer formulation is considered to be one of the potential strategies in order to improve the shelf-life of the biofertilizer. As compared to solid carrier based biofertilizers, liquid biofertilizer formulations allow its manufacturer to include sufficient amount of nutrients as well as cell protectant and inducers which are responsible for cell/spore/cyst formation in order to ensure prolonged shelf-life. Shelf-life of commonly used solid carrier based biofertilizers is about six months while it could be as higher as 2 years for liquid formulation. Other than this, solid carrier based biofertilizers are less thermo-tolerant as compared to liquid formulations which can tolerate the temperature as high as 55°C.

Hence, better shelf-life can be achieved by the application of a liquid biofertilizer formulation. Though, process cost of liquid biofertilizer is significantly higher as compared to solid formulation. So the successful commercialization of cheap liquid biofertilizer is grated challenge as shelf-life of these products is still a major concern in field of applications of biofertilizers technology.

Adoption of biofertilizers technology by farmers and major crops

The common claim that conversion to organic agriculture on large-scale may result in drastic reduction in globe food supplies as well as large increases in conversion of undisturbed lands to agriculture, has not been borne out in numerous modelling studies. Different conversion studies have showed that the consumption of domestic food would not suffer while the exports would alter depending on crop but the organization of farming would definitely modify with better diversification of agriculture. Huge conversion to the organic agriculture defiantly resulted in enhancement in crop yield over the current averages. A study in northern Europe revealed that demand of domestic food could be fulfilled with organic methods, but food would be more expensive in this way. Quite similar conclusions were also made in different studies which were conducted in USA. An important study conducted in Germany leads to the conclusion that if per capita dietary calories of meat were reduced to 21% from the current used 39 percent all German agricultural lands could be transformed to organic agriculture without any increase in existing imports or expansion of agricultural land. According to this report such a conversion are possible in veru near future as the survey highlighted that 40 percent of youth from Germany have planned to adopt low or no-meat diets. The threshold for this type of conversion may come very soon, as the trend of abandonment of meat-consumption is also increasing in Europe as a result of BSE33. Many other studies have also proved the benefits of having vegetarian diet over non-vegetarian diet in terms nutritional advantages as well as of energy consumed for food production. Organic farmers no doubt grow a diversity of crops and also maintain livestock in order to optimize the useage of nutrients as well as the space among species. This ensures several countless. But this can have very significant impact on local food security and flexibility. In rain-fed systems, organic agriculture has been illustrated to do better than conventional agricultural systems even under environmental stress conditions. Under the faceable conditions, the market returns from organic agriculture can significantly contribute to local food security and will definitely help in increasing family income. Around the world, especially in developing countries which bear high population pressure and with the current state of knowledge as well as technology, organic farmers are helpless in producing enough food for everybody.

Nitrogen fixers

The German scientists, Hellriegel and Wilfarth, in 1886, were the first ones to discover nitrogen fixation. For the first time they reported that legumes, having root nodules, are able to use gaseous (molecular) nitrogen. Much later, the process of nitrogen fixation in blue-green algae which are now classified as cyanobacteria was established. Since that time, research efforts in the field of nitrogen fixation have dramatically increased which results in the selection of several strains showing numerous beneficial features.

Nodule formation

Rhizobial infection which occurs via plant root hairs prior to the infection process; significantly respond to the occurrence of compatible rhizobia mostly by deformation. At this stage of deformation, the plant perceives the rhizobial signal first and then initiates a developmental program which aimed at development of symbiotically nitrogen-fixing nodules. A nodule meristem is thus formed within the root while the rhizobia enter through a plant derived infection thread-a tube formed to facilitate rhizobia entry to the deeper layers. The infection threads normally grow transcellularly and then rhizobia which are wrapped into a plant derived membrane, and now called "symbioses membrane", are delivered into plant cells.

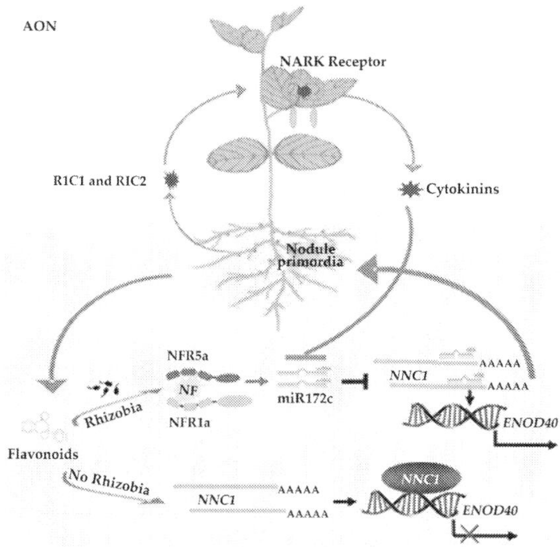

Figure: Nitrogen fixation-nodule formation (Source: Yinping Tian, Qiong Jiang, Brett J. Ferguson, Peter M. Gresshoff and Xia Li Youning Wang, Lixiang Wang, Yanmin Zou, Liang Chen, Zhaoming Cai, Senlei Zhang, Fang Zhao, Soybean miR172c Targets the Repressive AP2 Transcription Factor NNC1 to Activate *ENOD40* Expression and Regulate Nodule Initiation, The Plant Cell Preview, 1-20, 2014)

Nodule physiology

During formation of the nodule, host tissues tend to develop to form a specialized tissue which maintains such an environment in which nitrogen fixation can occur. In the specific nodule, specialized (organelle-like) forms of bacteria called "bacteroids" are engulfed into the plant-derived membranes, forming symbiosomes. Photosynthetic is then transported to the nodule in form of sucrose; however dicarboxylic acids additionally provide the bacteroids with carbon as well as energy through the symbioses membrane. Ammonia which produced in the bacteroid is required to be transported to the plant via the symbioses membrane. If a root nodule is cut to open and the inside of a root nodule is pink/red showing that the nodule is active and fixing lots of nitrogen for the plant. This color is due to the presence of leghemoglobin in large amount. The more redder the nodule indicates the more active it is. When nodules are very young and are not yet fixing nitrogen they are found to be white or grey inside. Legume nodules who are no longer fixing nitrogen automatically turn green and are discarded by the plant. This change in color is the result of an inefficient Rhizobium strain or poor plant nutrition. After harvest legume roots left in the soil decay, release organic nitrogen compounds for uptake by the coming generation of plants. Farmers take great advantage of this natural fertilization by rotating a leguminous crop with a non-legsuminous one.

Nitrogen fixation and crop productivity

For optimum and best plant growth, the availability of nutrients in sufficient and balanced quantities is very much important. After photosynthesis, the process of nitrogen fixation is the second most important process which play role in crop production. The process of photosynthesis is known to capture sunlight and then to produce energy while in process of nitrogen fixation nitrogen gas is used to form ammonium. Nitrogen fixation process can significantly provide up to 300-400 kg N/ha/yr for free. Annually, Biological nitrogen fixation process, signify up to 100 million tons of N for terrestrial ecosystems while from 30 to 300 million tons for marine ecosystems. It was the end of 19th century when the very first industrial production of rhizobium inoculant began. Since the Second World War the supply and demand of nitrogenous chemical fertilizers has been dramatically increasing in order to sustain and maintain production of (i) cereal crops, (ii) legumes and (iii) other plants of agricultural importance. Biofertilizers can significantly add 20-200kg N ha (by process of fixation), liberate huge amount of growth-promoting substances as well as can increase crop yield by 10-50%. They are not only cheaper, free of pollution, completely based on renewable energy sources as well as also improve soil quality. Both, plant and the bacterium, live separately but their association is very advantageous for both of them. It has been reported that in plants, up to 25 percent of total nitrogen directly came from nitrogen fixation. Activity of nitrogen fixing microorganisms greatly depends upon two things i.e., (i) excessive amount of carbon compounds and (ii) adequately low level of combined nitrogen. The most important and clear purposes of applications of bio fertilizers to soil are: (i) to provide nutrient sources as well as good soil conditions for the growths of crops (when used as a live body); (ii) to moderately substitute and improve the function of chemical fertilizer as well as control the application quantities of the fertilizers and ultimately to maintain the same crop yields. The capital which used for production of bio fertilizers is not only cheaper than that of chemical fertilizers yet also lessen the adverse effects aroused from the application of chemical fertilizers to soil. To the broader aspect, the indirect significant purposes of applying bio fertilizers to soil are

- to enhance the growth of root system in order to increase the water and nutrient absorption abilities of crops
- to extend the life of root
- to neutralize as well as degrade harmful materials accumulated in soil
- to promote survival efficiency of seedling after transplanting
- To get shorter time for the flower to come out.

Bio fertilizer and bio fungicide in food security

Biofertilizer is a "substance that contains living microorganism which, when applied to seed, plant surfaces, or soil, colonizes the rhizosphere or the interior of the plant and promotes growth by increasing the supply or availability of primary nutrients to the host plant". Biofertilizers are not typical fertilizers because fertilizers acts by directly increasing soil fertility by addition of nutrients, on the other hand biofertilizers add nutrients via natural processof atmospheric nitrogen fixation, solubilizing phosphorus and also by stimulating growth of targeted plant through the synthesis of several growth promoting substances.

Biofertilizers are widely used for application of seed, soil or composting areas with the obvious objective to enhance the numbers of such microorganisms as well as in order to accelerate certain microbial process to expand the extent of nutrients availability in a form which can be well assimilated by plant. Such bio fertilizers are potential components of integrated nutrients management in soil and they also play significant role in productivity and sustainability of soil. In the modern era, the bio fertilizer is rapidly replacing chemical fertilizers as they are cost effective, ecofriendly and also renewable source of plant nutrients.

Due to plant diseases there in significant trend in yield reductions which is estimated to be almost 20 % in agricultural as well as horticultural crops worldwide. With the breakthrough in discovery of synthetic fungicides a greater increase in food production and control on diseases is observed. But the other side of coin revealed that there diverse applications has raised a number of ecological and environmental issues. According to CGIAR, it is only sustainable agriculture which offers the successful management of resources inorder to satisfy the changing human needs as well as maintaining the quality of environment and conserving natural resources as well. The clear choice for that will be the adaptation of bio-fungicide technology in order to combat plant diseases, as they are comparatively safe, biodegradable as well as eco-friendly. Bio-fungicides exhibit various mechanisms like (i) antibiosis, (ii) competition, (iii) mycoparasitism and (iv) induce systemic resistance to fight against harmful plant pathogens. The simplified and convenient registration system of biopesticides in recent years also allowed commercial pilot production of bio-fungicides in various countries. Therefore, many commercial formulations are now easily available in the market. Development of microbial consortium will greatly improve disease control via synergy in signal cross talk between microbes. Genetically improved antagonistic microorganisms tend to increase their effectiveness as bio-fungicides with an increased antifungal activity. Application of genetically manipulated bio-fungicides in modern agriculture offers tremendous opportunities for expanding food production, lowering risks in food production, improving environmental protection as well as strengthening food marketing and enhancing the livelihood of farmers and thus ensuring food security in many developing countries.

Figure: Role of Biofertilizers in food security (source: http://www.garden-salive.com/product/soap-shield-flowable-liquid-copper-fungicide-3/lawn-care)

Mechanism involved in plant growth improvement

Many rhizospheric bacterial strains possess plant growth-promoting mechanisms. These bacteria can be applied as biofertilizers in agriculture and forestry, enhancing crop yields. Bacterial biofertilizers can improve plant growth through several different mechanisms:

- The synthesis of plant nutrients or phytohormones, which can be absorbed by plants,
- The mobilization of soil compounds, making them available for the plant to be used as nutrients,
- The protection of plants under stressful conditions, thereby counteracting the negative impacts of stress, or
- Defense against plant pathogens, reducing plant diseases or death.

Several plant growth-promoting rhizobacteria (PGPR) have been used worldwide for many years as biofertilizers, contributing to increasing crop yields and soil fertility and hence having the potential to contribute to more sustainable agriculture and forestry. The technologies for the production and application of bacterial inocula are under constant development and improvement and the bacterial-based biofertilizer market is growing steadily. Nevertheless, the production and application of these products is heterogeneous among the different countries in the world. Basically, biofertilizers are carrier based preparations containing specific strains of microorganisms like bacteria, fungi, algae or their combinations.

Potential of biofertilizer to replace chemical fertilizer

The use of chemical fertilizers (e.g. urea, calcium nitrate, ammonium sulphate, diammonium phosphate etc.) have a great importance for the world's food production as it works as a fast food for plants causing them to grow more rapidly and efficiently. While adverse effects are being noticed due to the excessive and imbalanced use of these synthetic inputs. Moreover, persistent use of conventional chemical fertilizers subverts the soil ecology, disrupt environment, degrade soil fertility and consequently shows harmful effects on human health and contaminates ground water. For these reasons, biofertilizers, the organic substances, which make use of microorganisms to increase the fertility of soil, has been identified as harmless input help in safeguarding the soil health and also the quality of crop products. Biofertilizers add nutrients through the natural processes of nitrogen fixation, solubilizing phosphorus, and stimulating plant growth through the synthesis of growth promoting substances. They are also environment friendly and responsible for continuous availability of nutrients from natural sources.

Table: Different features of chemical fertilizers and biological fertilizers

Features	**Chemical Fertilizers**	**Biofertilizers**
Raw material	Non-renewable	Renewable
Energy	Fossil Fuels	Solar
Catalyst	Al, Fe, Mo Oxides	Nitrogenase enzymes
Reductant	H_2	Organic
Temperature and Pressure	750◦F, 200-600atm	Ambient, T, P
Energy Requirement	680KJ.mol^{-1}.NH_4^+	355 KJ.mol^{-1}.NH_4^+
Efficiency	40-45%	90%
Pollution Effetect	Pollution exist due to indiscriminate use	No pollution.
Soil Health	Detoriates	Improves

2
CHAPTER

BIO-FERTILIZERS: TYPES AND IMPORTANCE OF BIO-FERTILIZERS

Bio-fertilizers: Types and Importance

Chemical fertilizers are mostly used in the large amounts for the purpose to enhance output in various product varieties of crop plants. As, chemical fertilizers are the biggest reason of posing pollution to the water bodies as well as ground water, which (contamination) also getting bioaccumilate in crop plants hence badly destroying ecological cycles. Therefore, modern scientists with a dream to encourage sustainable agriculture are focusing for switch completely from chemical farming to organic farming. In organic farming crops are not polluted as they rose through use of manures, bifertilizers and biopesticides which provide optimum nutrients to plants, keeping pests and pathogens in control.

Biofertilizers are actually "microorganisms which bring roughly nutrient enrichment of soil through enhancing the availability of nutrients to plant crops". These micro-organisms who exhibit the ability to act as bio-fertilizers are the bacteria, cyanobacteria and mycorrhizal fungi. Bacteria as well as cynobacteria exhibit the characteristic of nitrogen fixation whereas mycorrhizal fungi preferentially withdraw minerals out of organic matter for the plant with which they are related. Nitrogen fixation is "the process of conversion of molecular or dinitrogen into nitrogen compounds". Insoluble forms of soil phosphorus are transformed into soluble forms by action of specific micro-organisms. This makes the phosphorus readily available to the plants. Phosphate is also solubilised by some bacteria and by some fungi as well that build up association with plant roots.

Types of Biofertilizers

The various types of biofertilizers which help the plant to grow at different levels of its growth are given in figure below:

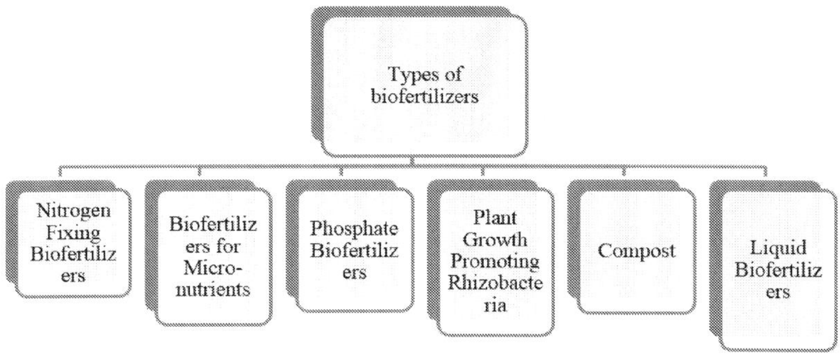

Figure: Different types of Biofertilizers

The sub classification of the above described types is:

Nitrogen Fixing Biofertilizers

Increase soil nitrogen level ii. Fixes the atmospheric nitrogen in the soil and make it available to the plants

Examples: *Azotobacter*, *Nostoc*, *Rhizobium*, *Azospirillum*

a) Azobacter
   - Free living N-fixing bacterium
   - Non-leguminous plants

b) Cyanobacteria
   - Nostoc/Blue green algae
   - Free-living as well as symbiotic
   - Fix atmospheric nitrogen to soil

c) Rhizobium

- Symbiotic N-fixing bacterium
- Leguminous plants

d) Azospirillum

- Associative symbiotic N-fixing bacteriumii. Graminaceous plants

2. *Phosphate Biofertilizers*

- Phosphorous Solubilizing Biofertilizersb. Phosphorus Mobilizing Biofertilizers

a) Phosphorus Solubilizing Biofertilizers

- Solubilize the insoluble phosphate from organic and inorganic phosphate sources
- Releases insoluble phosphorus in soil and fix in clay minerals
- Secrete organic acids and lower the pH to dissolve bound phosphates in soil

Examples: Species of *Bacillus*, *Pseudomonas*, *Penicillium*, *Aspergillus*

b) Phosphorus Mobilising Biofertilizers

Transfer phosphorus from the soil to the root cortex

Examples: Arbuscular Mycorrhiza (AM fungi)

i. Fungus penetrates the cortical cells of the roots

Increase surface area of roots

iii. Displace of absorption equilibrium of phosphate ions which increases the transfer of P ions

- Stimulate metabolic processes
- Arbuscles absorb these nutrients into the root system

3. *Biofertilizers for Micro-nutrients*

Bacterial species are Silicate and Zinc solubilizers

- Degrade silicates and aluminium silicates in soil
- Help in silicate weathering
- Example: *Bacillus* sp.

4. Plant Growth Promoting Rhizobacteria (PGPRs)

- Act as both Biofertilizer and Biopesticides
- Promote growth by:
- Improved Nutrient Availability (Biofertilizers)
- Suppression of Plant Disease (Bioprotectants)
- Phytohormones Production (Biostimulants)

5. Compost

6. Liquid Biofertilizers

Brief description of these types is given in this chatter while their details will be discussed in coming sections.

Free Living Nitrogen Fixing Bacteria

Free Living Nitrogen Fixing Bacteria live freely in the soil and perform the function of nitrogen fixation. Few among them are saprotrophic and are living on organic remains, e.g., Azoto- bacter, Bacillus polymyxa, Clostridium, Beijerinckia. They can further be distinguished into aerobic and anaerobic forms. This particular property to fix nitrogen is also exhibited by photoautotrophic bacteria, e.g., *Rhodopseudomonas*, *Rhodospirillum* as well as *Chromatium*. Inoculation of soil with these bacteria assists in enhancing yield and thus reducing of nitrogen fertilizers use. *For example*, Azotobacter when occured in fields of Cotton, Maize, Jowar and Rice, not only enhances yield yet also saves nitrogen fertilizer to the tune of 10-25 kg/ha. Its inoculation is accessible under the trade name of azotobactrin. Azospirilla are Gram-negative free-living nitrogen-fixing rhizosphere bacteria. They display a flexible C- and N-metabolism, which makes them very well adapted in order to establish in the competitive environment of the rhizosphere. Ammonium, nitrate, nitrite, amino acids and molecular nitrogen can serve as N-sources.

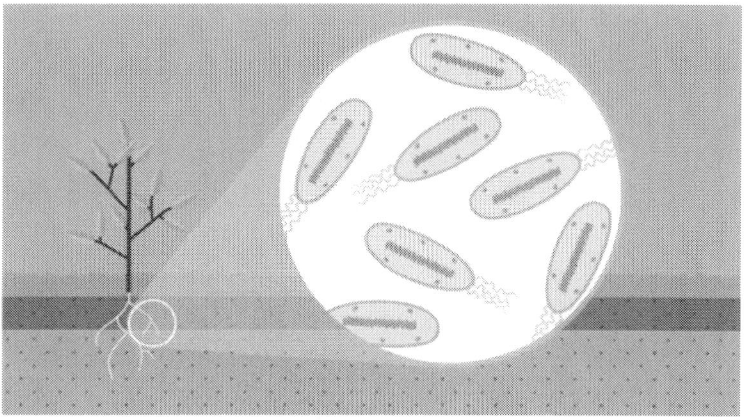

Figure: Free Living Nitrogen Fixing Bacteria (source: http://www.bbc.co.uk/education/guides/z72v4wx/revision/2)

Free Living Nitrogen Fixing Cyanobacteria:

Majority of free living cyanobacteria or blue-green algae have the property of nitrogen fixation among them are included Anabaena, Nostoc, Aulosira, Totypothrix, Cylindrospermum, Stigonema. Cyanobacteria are photosynthetic. Due to this property they put in organic matter as well as extra nitrogen to the soil.

Aulosira fertilissima is considered to be one of the most active nitrogen fixer of rice fields. Cylindrospermum licheniforme usually grows in Sugarcane and Maize fields. Cyanobacteria is another cheap i.e., low price biofertilizer. In few developing countries, the technique and protocols of cyanobacteria inoculation to rice fields is already being adopted and implimented. Phosphate, Molybdenum and Potassium are also supplied additionally.

It was demonstrated in previous studies that free-living cyanobacteria and particularly legumes corresponds to immediate sources of N for the ecosystem in comparison to mosses and lichens, which release N via slow decomposition and upon disturbances. In N- limited arctic ecosystems, the co-occurrence of different N_2 fixers may thus correspond to a constant supply of 'new' N for the ecosystem by accompanig each other at time of N release.

Loose Association of Nitrogen Fixing Bacteria

Various nitrogen fixing bacteria like Azospirillum live around the roots of higher plants without developing any close relationship. This type of associating is mostly said to be rhizosphere association. The bacteria acquire some plant exudate and keep using the same to be part of their food necessity. The bacteria fix nitrogen and exude a part of the fixed nitrogen to be used by the plant. This prominient phenomenon is called as "associative mutualism" or "associative symbiosis".

Symbiotic Nitrogen Fixing Bacteria

Since past few decades, large number of researches has been conducted on the genetics, biochemistry, physiology, ecology, and agronomics of the bacteria forming nitrogen-fixing symbioses with legumes. These bacteria, collectively termed as to be the rhizobia, are taxonomically and physiologically assorted members of the α and β subclasses of the Proteobacteria. They mostly comprise members of the genera Rhizobium, Bradyrhizobium, Mesorhizobium, Sinorhizobium, and Azorhizobium.

They form a mutually favorable association with the plants. The bacteria acquire food and shelter from plants. While in return, they provide a part of their fixed nitrogen to the plants. The mainly significant of the symbiotic nitrogen fixing bacteria is Rhizobium (pi Rhizobia). It played role of formation of nodules on the roots of legume plants. There are about one dozen species of Rhizobium which build up association with variety of legume roots, e.g., R. leguminosarum, R. lupini, R. trifolii, R. meliloti, R. phaseoli.

These bacteria, also said to be rhizobia, are free living in the soil and are not able to fix nitrogen except for a strain of Cowpea Rhizobium. They only develop the skill to fix nitrogen when they occurred inside the root nodules. In the nodule cells, bacteria (bacteroids) lie in groups surrounded by membrane of the host which is lined by a pink-red pigment called leghaemoglobin. Presently cultures of Rhizobium specific for different crops are raised in the laboratory.

Frankia, nitrogen fixing mycelial bacterium (actinomycete), is strongly associated symbiotically with the root nodules of various nonlegume plants like Casuarina, Alnus (Alder) Myrica, Rubus etc.

Leaves of a few plants (e.g., Ardisia) develop special internal cavities for providing space to symbiotic nitrogen fixing bacteria, Xanthomonas and Mycobacterium. Such leaves are a continuous source of nitrogen fertilizer for the soil.

Symbiotic Nitrogen Fixing Cyanobacteria:

Nitrogen fixing cyanobacteria (blue- green algae) establish symbiotic association with variety of crops, e.g., cycad roots, lichens, liverworts, Azolla (fern). Out of these, Azolla-Anabaena association is of vital significance to agriculture.

Azolla pinnata is a tiny free floating fresh water fern which exhibit the ability to multiply rapidly i.e., doubling every 5-7 days. The fern may coexist with rice plants as it does not interfere with their growth. In some South-East Asian countries, especially China, the rice fields are frequently supplied with Azolla.

Anabaena azollae exist in in the leaf cavities of the fern. It fixes nitrogen. A portion of the fixed nitrogen is mostly excreted in the cavities and becomes available to the fern. The decaying fern plants release the same for utilization of the rice plants. When field is dried at the time of harvesting, the fern functions as the green manure, decomposing and enriching the field for the next crop.

Rhizobium

Rhizobia are soil bacteria that live on the roots of legumes by establishing on the root nodules and fix nitrogen therefore it plays an important role in agriculture and has a symbiotic relationship with the leguminous plants such as rice, wheat, maize, barley, millets, alfalfa, beans, peas and clover. They have rod shape, do not produce spores, and are motile and aerobic. It has been studied that their inoculation increases the yield of different cereal grains. It enhanced stomatal conductance which increased 12% photosynthetic rate and 16% grain yield in rice varieties. Likewise, an increase in photosynthetic rate and yield was also recorded in Bananas. It is suggested that Liquid inocula of Rhizobium should be developed for easy handling and spray. "Rhizobium bacteria are capable of fixing atmospheric nitrogen in association with leguminous crops. Rhizobium species enter the roots of host plants and form nodules on the root surface". "Rhizobium supplies nitrogen to the host plant, nitrogen fixed by the rhizobium is trans located through xylem vessels of the host plant mainly in the form of asparagine and to some extent as glutamine"." Rhizobium is a host species and different crops have different rhizobium species".

In a study, Congo red manitol medium was used to find the effect of Rhizobium on growth of sasbenia plant. It was separated out from nodules of roots of sasbenia plant and added on Cong Red Manitol medium, which developed two colored colonies on plates. Growth of Rhizobia was shown via white colonies at 4°C. Before plantation of the plants, physico-chemical properties of soil were analyzed. The results showed that pH was 7.9, moisture content was 11.1 %, total nitrogen 0.0945%, and phosphorus 7.30 ppm. The control group showed a mean value of 0.549 gm for fresh shoot weight, 4.16 gm root fresh weight, 0.0321 cm root heights, 16.66 cm Shoot height and number of branches were eight. The plants of experimental group showed the following results; 1.1700 gm Shoot fresh weight, 6.5 gm root fresh weight, 0.0541 cm root heights, 28.83 cm shoot height and number of branches were 11.66. Therefore it was concluded that inculcation of soil with Rhizbia not only improves production of plants but also helps in reduction of environmental burden from chemical fertilizers.

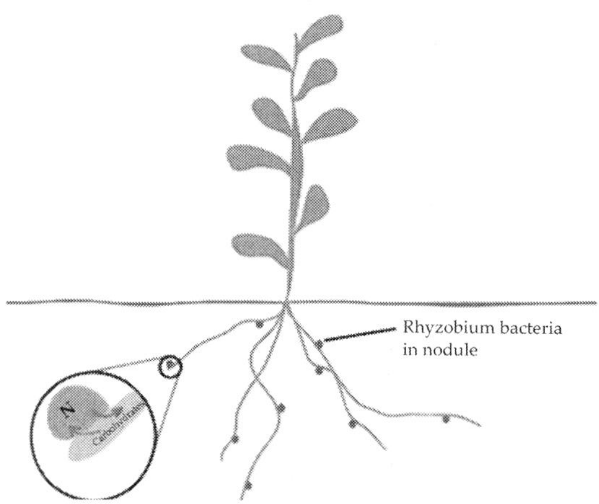

Figure: Rhizobium – biofertiler (source: https://www.permaculturereflections.com/rhizobium-symbiosis-with-woody-plants-leguminous-nitrogen-fixing-trees/)

Azotobacter

Azotobacter species play important roles in agriculture as it helps in atmospheric nitrogen fixation, production of phyto-hormones, degradation of toxic compounds and driving the ecological balance in agro-ecosystems. They normally do not have symbiotic relationship with plants. They have a range of enzymes that perform the nitrogen fixation including ferredoxin, hydrogenase, and nitrogenase. These synthesize biologically active substances like phyto-hormones such as auxins which stimulate plant growth. They enhance the mobility of heavy metals like cadmium, mercury and lead in the soil which bioremediate afterwards. Some species of Azotobacter biodegrade aromatic compounds that contain chlorine, such as 2,4,6-trichlorophenol, which was previously used as an insecticide, fungicide, and herbicide, but now it is a known mutagenic and carcinogenic. Soil samples from different zones of Colombia were collected randomly from different vegetable organic cultures with neutral pH in order to analyze Azotobacter's efficiency. It was found out that it also increases germination of seeds. Seeds having less germinating percent if inoculated can increase germination by 20-30%.

It is less effective in soils having poor organic matter. It can survive even in alkaline soils. They are tolerant to high salt concentrations and is the heaviest breathing organism which requires a large amount of organic carbon for its growth. Gerlach and Vogel (1902) initiated work on artificial inoculation of seeds with A. chroococcum and reported an increase of dry matter in buck wheat by 42 per cent. The use of Azotobacter to improve the growth of agricultural plants and soil properties is recommended in various studies. A. chroococcum and A. vinelandii have long been used as soil and seed inoculants. Azotobacter is a broad spectrum biofertilizer and can be used as inoculant for most of agricultural crops. With emphasis on development of sustainability in agriculture, Azotobacter is an important bioinoculant especially in organic farming. Azotobacter is also known to suppress phytopathogens or reduce their deleterious effects thereby improving plant growth. Incidence of fungal, bacterial and viral diseases in the crops is reduced by Azotobacter inoculation.

Methods of Application

There are three methods of application.

1. Seed treatment

Normally, 500 g of bio-fertilizers is required for seeds to be applied in a hectare of field. Bio-fertilizer is mixed with water and adhesive material so that seed coat is not broken, is dried for half an hour before sowing.

2. Seedling treatment

Bio-fertilizers are being used for seedling treatment. This is mainly done in transplanted crops. A slurry of Azotobacter biofertilizer is prepared. Seedlings are dipped in this slurry for about 15 minutes, allowed to dry and then transplanted.

3. Soil application

About 2 kg of bio-fertilizers are mixed with 40-50 kg of decomposed FYM and are broad cast at the time of sowing/prior to sowing.

Azospirillum

Azospirillum, a rhizobacteria, also exhibit the property to promote plant growth, fixes nitrogen and produces phytohormones. It is known to be safest bacteria for plants, environment as well as for hum a beings. It is frequently used for cereals and grasses particularly for wheat and rice. They are bacteria with high salt tolerant rate with the strong potential to solubilise phosphate. Five species of Azospirillum have been explained to date which includes *A. brasilense*, *A.lipoferum*, *A.amazonense*, *A.halopraeferens* and *A.irakense*. The organism proliferates under both anaerobic and aerobic conditions. Apart from nitrogen fixation, growth promoting substance production, disease resistance and drought tolerance are few of the supplementary benefits due to Azospirillum inoculation.

Methods of Application

The mixture of Azospirillum, Azotobacter (Gmax Nitromax) and phosphate solublizing bacteria (Gmax Phosphomax) can be used for non-leguminous crops-like cereals, oilseeds cotton, vegetable and all other crops which are usually planted as seeds.

Seed treatment:

For one kg of seeds about 40 grams of Gmax Nitromax product is required. Mix required quantity of Gmax Nitromax with equal quantity of Rice gruel. With this mixture, mix the required amount of seeds shade dry and after 24 hours, use the seeds of sowing.

Seedling Root dip:

Mix 2 Kg of Gmax Nitromax with 50 lits of water. In this mixture, keep the seedling roots in immersed condition for 10 minutes and use the seedlings for transplantation in the field.

Soil application:

Mix about 5 kgs Gmax Nitromax with 100 kgs of organic manure, keep the mixture under shade for one week and apply to the soil.

Cyanobacteria

Cyanobacteria is a blue green algae that uses solar energy, nitrogen and water in order to promote soil fertility and growth of plant. These are also nitrogen fixing bacteria that are both economically and ecologically attractive. When it was inoculated to the plants of Ethopia, the yield was found to be more than the plants treated with urea. Nutritional values of these plants were also increased as zinc was more than double and iron more than fifty percent as compared to the crops treated with Urea [8]. It also improves physio-chemical characteristics of soil like moisture retaining capacity and nutrient concentrations. Cyanobacteria have also been genetically modified as they are being used as biofuels for example biodiesel, syngas, and biomethane. Cyanobacterial biofertilizers were used for rice crops in India for research purpose which indicated a strong potential for cyanobacterial biofertilizer technology in rice-growing countries.

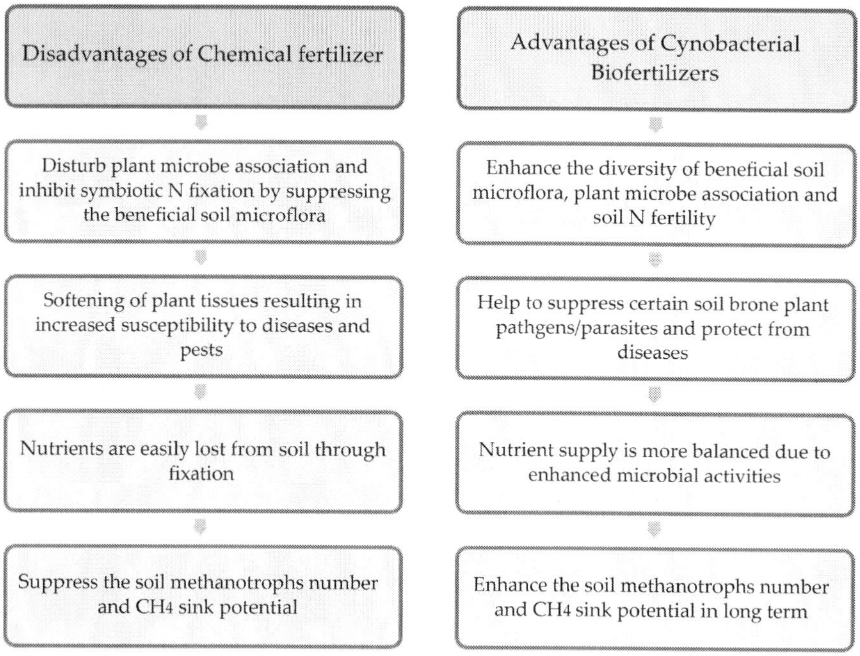

Figure: Chemical fertilizers vs Cynobacterial Biofertilizers

Azolla

Azolla, a free-floating water fern, floats in water and have the ability to fix atmospheric nitrogen in association with nitrogen fixing blue green alga Anabaena azollae. Azolla is a potential biofertilizer to be used for wetland rice and it is also known to supply 40-60 kg N/ha per rice crop. In past, before its cultivation as a green and ecofriendly manure, Azolla was utilized as a fodder for domesticated animals such as ducks. In recent era, Azolla is widely used as a sustainable and ecofriendly feed substitute for livestock particularly dairy cattle, poultry, piggery and fish. Azolla contains 25 – 35 per cent protein on dry weight basis and enriched in essential amino acids, minerals, vitamins and carotenoids. Azolla is a vital cheap (low cost input) which holds the ability to play key role in enhancing soil quantity in sustainable rice farming. The dual potentials as biofertilizer as well as animal feed make this, water fern, Azolla an efficient input to both the vital components of integrated farming, agricultural as well as animalo husbandry.

Azolla, a water fern, needs a temperature of 35-38° C in order to grow. The multiplication of Azolla is strongly affected under elevated temperature. In orders to examine the feasibility of using Azolla in flooded rice fields, a study was conducted as a series of experiments in PVC tanks during 2000–2002 in Po Valley (northern Italy), to research the growth, development as well as the resistance/tolerance to low temperatures and to usually used herbicides of numerous Azolla strains. Three among five strains which were tested survived the winter, with enhancement in biomass producing approximately 30–40 kg per hectare of nitrogen.

Microphos Biofertilizers

Various soil micro-organisms exhibit vital role in transformation of nutrients for agricultural use. Few of them are capable of fixing nitrogen, while other may enhance the availability of nitrogen and phosphorus . Bio-fertilizers are "the products containing living cells of micro-organisms that have the ability to mobilize the nutrients from non-usable form through biological processes". Bio-fertilizers also referred to as "microbial inoculants" bears great potential to act as supplementary, renewable and ecofriendly source of plant nutrients and are vital component of Integrated Plant Nutrient System (IPNS) .

Plant nutrients are lost to the environment from soil in several ways. Their large quantities are removed from the soil when crops are harvested, other than these weeds also remove a large amount of plant nutrient from soil. Nutrients are also get removed by leaching and erosion . Nitrogen is mostly removed from plants by volatilization and de-nitrification. In order to increase production, preserve soil health, minimize nutrient losses, improve soil environment and reduce energy consumption, the most important thing is the use bio-fertilizers along with chemical fertilizers . Bio-fertilizers will assist in fixing atmospheric nitrogen, dissolve soil phosphorus and stimulate plant growth through synthesis of growth promoting substances.

Latest soil management protocols and procedures are mostly dependent on inorganic chemical-based fertilizers, which pose a serious risk to human health as well as interacting environment. The exploitation of useful microbes as a biofertilizer is now of supreme significance in agriculture sector for their potent role in food safety and sustainable crop production. The eco-friendly approaches inspire a broad range of application of plant growth promoting rhizobacteria (PGPRs), endo- and ectomycorrhizal fungi, cyanobacteria and many other useful microscopic organisms led to enhanced nutrient uptake, plant growth and plant tolerance to abiotic and biotic stress.

Mycorrhiza

"Organic farming" is greatly dependent on the natural microflora of the soil which represents all kinds of useful bacteria and fungi taking into count the arbuscular mycorrhiza fungi (AMF) usually termed as "plant growth promoting rhizobacteria (PGPR)". Biofertilizers bears the ability to keep the soil environment enriched with various kinds of micro- and macro-nutrients via nitrogen fixation, phosphate and potassium solubalisation or mineralization, release of plant growth regulating substances, production of antibiotics and biodegradation of organic matter in the soil.

- Ectomycorrhiza
- Endomycorrhiza

Ectomycorrhiza

Ectomycorrhizal (ECM) fungi have been shown to improve growth of the host plant before the formation of physical ECM structures, i.e. during the so-called pre-mycorrhizal phase. Formulation of the Ectomycorrhiza between filamentous fungi as well as forest tree roots is vital in order to satisfy the nutritional demands of main forest trees. Ectomycorrhizas as well as the symbiotic fungal "mycelia" which is mainly termed as the "mycorrhizosphere" are in strong association with varied bacterial communities. Only few studies on the fundamental consequences of EM associated bacteria have been conducted uptill now. Yet, their diversified roles have already been demonstrated for these bacteria. In environment which is not adapted well to fire, post-fire tree employment is frequently sparse. Huge quantities of carbon are exchanged between terrestrial ecosystems and the atmosphere, and extensive research efforts are made to understand carbon cycling and the impact of elevated atmospheric CO_2 levels. The response of soils to increased carbon availability is largely driven by root associated ectomycorrhizal fungi in forest ecosystems, since they partition host derived carbon belowground. The fungal symbionts forming ectomycorrhizas, as well as their associated bacteria, benefit forest trees in a number of ways although the most important is enhancing soil nutrient mobilization and uptake. This is reciprocated by the allocation of carbohydrates by the tree to the fungus through the root interface, making the relationship a mutualistic association. Many field observations suggest that ectomycorrhizal fungi contribute to a number of key ecosystem functions such as carbon cycling, nutrient mobilization from soil organic matter, and nutrient mobilization from soil minerals and linking trees through common mycorrhizal networks. Until now, it has been very difficult to study trees and their fungal associates in forest ecosystems and most of the work on ECM functioning has been done in laboratory or nursery conditions.

Endomycorrhiza

An Endomycorrhiza arbuscular mycorrhiza (plural mycorrhizae or mycorrhizas) is a type of mycorrhiza in which the fungus penetrates the cortical cells of the roots of a vascular plant". "rbuscular mycorrhizae (AMs) are characterized by the formation of unique structures such as arbuscules and vesicles by fungi of the phylum Glomeromycota (AM fungi).AM fungi help plants to capture nutrients such as phosphorus and micronutrients from the soil. It is believed that the development of the arbuscular mycorrhizal symbiosis played a crucial role in the initial colonization of land by plants and in the evolution of the vascular plants.

BENEFITS OF ENDOMYCORRHIZAE

The arbuscular endomycorrhizae are considered to be ecologically important for vascular plants in view of their beneficial effects on plant growth and survival. They are one of the few plant fungus associations with a fossil record and may even have facilitated the origin of land flora. The enhanced growth of plants infected by vesicular–arbuscular (VA) mycorrhizal fungi results primarily from improved uptake of soil phosphate. Extra phosphate reaches the root through the fungal hyphae, which tap the soluble P in soil beyond the phosphate-depletion zone near the root surface. This mechanism, can explain the many correlations between root geometry and mycotrophy and other nutritional effects of VA mycorrhizae such as increased uptake of zinc and copper ions. The presence of VA endomycorrhizae has several benefits for the host plant.

They have an increased ability to absorb phosphorous and other elements.

Nitrogen Biofertilizers

Bio-fertilizers are micro-organisms which bring about nutrient enrichment of soil by enhancing the availability of nutrients to crops. The micro-organisms which act as bio-fertilizers are bacteria, cyanobacteria (blue green algae) and mycorrhizal fungi. Bacteria and cynobacteria have the property of nitrogen fixation while mycorrhizal fungi preferentially withdraw minerals from organic matter for the plant with which they are associated. Nitrogen fixation is the process of conversion of molecular or dinitrogen into nitrogen compounds. Insoluble forms of soil phosphorus are converted into soluble forms by certain micro-organisms. This makes the phosphorus available to the plants. Phosphate is also solubilized by some bacteria and by some fungi that form association with plant roots.

Arbuscular Mycorrhizal Fungi and its role in the Ecosystem

Arbuscular Mycorrhizal Fungi (AMF) constitutes a group of root obligate biotrophs that exchange mutual benefits with about 80% of plants. They are considered natural biofertilizers, since they provide the host with water, nutrients, and pathogen protection, in exchange for photosynthetic products. Thus, AMF are primary biotic soil components which, when missing or impoverished, can lead to a less efficient ecosystem functioning. The process of re-establishing the natural level of AMF richness can represent a valid alternative to conventional fertilization practices, with a view to sustainable agriculture. They provide the host plant with mineral nutrients and water, in exchange for photosynthetic products. The AMF mycelium which arises from the root system may acquire nutrients from soil volumes that are unreachables to roots". Furthermore, fungal hyphae are much thinner than roots and are therefore able to penetrate smaller pores. Carbohydrates and mineral nutrients are then exchanged inside the roots across the interface between the plant and the fungus. AM fungal hyphae entirely colonize the root cortex and form extremely branched structures inside the cells, i.e., arbuscusles, which are strongly considered as the functional site of nutrient exchange. Thus, AMF can improve the limitation in growth of plant occurred due to an inadequate nutrient supply. Other than an improved nutritional supply, AM interactions also provide various potential benefits to plants including improved drought and salinity tolerance and disease resistance.

Metals like Fe, Cu and Zn play significant roles in several sub cellular compartments. But these metals constitute a strongly reactive group of various elements which are toxic at higher concentrations. AM fungi are known to ease heavy metal toxicity in their host plants as well as to tolerate higher metal concentrations in the soil. Moreover, AM fungi can also pose a direct effect on the ecosystem, as they improve the soil structure/ aggregation AM fungi played a key role in reducing emissions of N_2O, an important greenhouse gas, so they could function in the mitigation of climate change. AM fungi also control N_2O emissions by enhancing plant N uptake and assimilation, which results in the reduction of soluble N in the soil, and, consequently, in a limitation of denitrification. Furthermore, the correlation between the key genes involved in N_2O production (nirK) and consumption (nosZ) and AM fungal abundance suggests that the regulation of N_2O emissions is caused by changes induced by AM fungi in the soil microbial biomass and in the community composition. AM fungi also have indirect pressure on GHG emissions and have the ability to change the physical conditions of soil, i.e., moisture, aggregation, and aeration, all of which influence the production and transport of GHG in soil. AM symbiosis helps to regulate N_2O emissions at high soil moisture levels and suggested that the control of N_2O emissions by AM plants could be driven by a higher use of soil water rather than by increased N uptake. Thus, AM fungi are primary biotic soil components which, when impoverished, e.g., due to anthropic input, can lead to a less efficient ecosystem functioning. The procedure of re-establishing the natural level of AMF richness can represent a promising alternative to conventional fertilization practices, with a view to ecofriendly as well as sustainable agriculture, a key target for growers facing the global recession and having great potential to deal with a more environmentally aware clientele.

PHOSPHOROUS BIOFERTILIZER

Main function

Phosphorous biofertilizers are that type of biofertilizers which have the ability to solubilize the phosphate and are present in the soil in an insoluble form. They greatly help in releasing that phosphate into the soil.

Microbes involved in phosphorous biofertilizers

The commonly used phosphate biofertilizers are *Bacillus species, Pseudomonas, Aspergillus* and *Penicillium.*

They are basically bacteria and fungi which are responsible for providing phosphorus to the plants.

Other way of forming phosphorous biofertilizer

The phosphorous biofertilizer can be obtained from these sources

Municipal sludge

The municipal waste water treatment sludge has increased content of dissolved phosphorous.

Ash from incinerators

The ash of municipal sludge can be utilized as a phosphorous biofertilizer but the problem associated with it is it contains toxic materials.

Bones

Currently the European Union has banned the use of bones and meat in the animal fodder which ends up in landfills, this can be best utilized as the biofertilizer and can reduce the burden on the land. But unprocessed raw material has low availability of phosphorous to plants.

Advantages

- clay minerals: They help in fixing the clay minerals by enhancing the soil microbes' activity which helps in the fixation of soil components.
- ***Lowering the pH of soil to dissolve the bound phosphates*:** They basically secrete the organic acids which lowers the pH of the soil and releases the bound phosphate from the soil.
- ***Helps in increasing crop yield:*** It helps in increasing the crop yield up to 200 to 500 kilograms/hectors.

- ***Helps in increasing Plant height:*** The phosphorous biofertilizer can significantly increase the plant height. In a research conducted on Marigold the application of phosphorous biofertilizer increase its height from 24.00 cm to 25.59 cm by applying the phosphorous biofertilizer on the roots the seeds of the marigold.
- ***Helps in increasing leaves:*** The phosphorous biofertilizer can increase the number of leaves in a plant. According to a research conducted in Iran on Marigold the phosphorous biofertilizer tends to increase the leaves on the pant from minimum 29.32 leaf per plant to the 39.06 per plant .
- ***Helps in increasing flower diameter:*** The phosphorous biofertilizer application on the Marigold plant can flower diameter from 57.118 mm to 68.871 mm. This can help in help plant in increased pollination as it become a little bigger.
- ***Regulate plant cell functions:*** The phosphorous biofertilizers performs several functions for plant cell by providing the phosphorus needed for phospholipids, gene expressions, metabolism, enzyme activity and development of roots.

SOLUBILIZING BIOFERTILIZERS

Main function

Nitrogen and phosphorous are the main nutrients needed mostly by the plants for their growth, metabolism and other processes. But unfortunately, the phosphorous is the least available nutrient for the plant in soil. Phosphorous is also less mobile which makes it harder for the plants to take it up easily. The phosphatic fertilizers are very costly that's why phosphate solubilizing biofertilizers are used which helps in the dissolution of the bound phosphates present in the soil. They convert insoluble phosphate $HPO_4^{2-}$to soluble phosphate $H_2PO_4^-$.

Microbes involved in solubility biofertilizers

Phosphate solubilizing biofertilizers are *Agrobacterium, Trichoderm, Enterobacter, Aspergillus, Pseudomonas, Glomus* and *Bacillus*. They are applied around the roots of the plants.

Preparation of solubilizing biofertilizer

The compost is mixed with the fungi or bacterial culture of solubilizing microbes at initial 60-65% moisture in a plastic container. The material is turned over every 3 to 4 days for about 56 days with variations in the temperature and pH. The pH tends to rise at about 8th week and decreases afterwards because of the depletion of the organic matter. During initial phases of composting (39.5%) and next two days the moisture was raised (57.5%) be nearly 18% and dropped in next two days. It went up by 23% again in 2 days and drastically lost 33% of moisture from 12th day onwards.

Advantages

- ***Protection from pathogen:*** The phosphorus released due to the solubility biofertilizers, helps in keeping the plant healthy which helps the plant protected from the pathogens.
- ***Offset phosphatic fertilizers:*** The use of phosphate solubility biofertilizers can used instead of phosphatic fertilizers to reduce the cost.

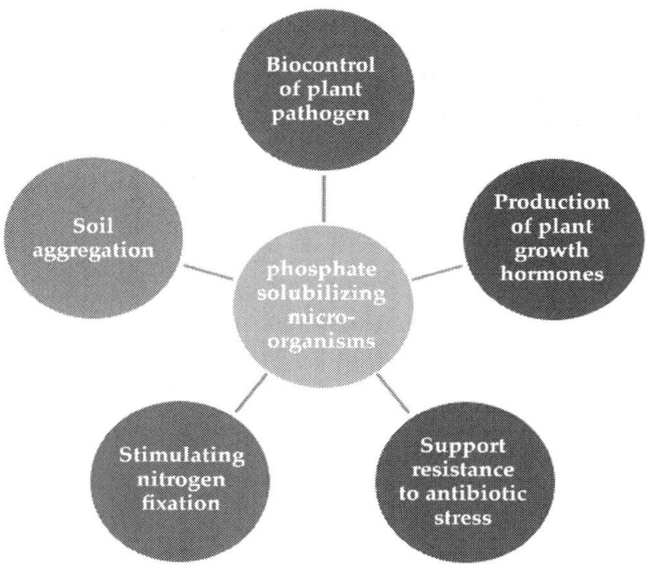

Figure: Role of phosphate solubilizing bacteria in plant growth and development

MOBILIZING BIOFERTILIZERS

Main function

The nutrients available in soil may not be available to the plants every time. They are sometimes bound with other compounds, which make even more harder for the plants to take up the minerals through roots. The mobilizing biofertilizers helps in mobilizing the minerals from their sources and make them available to the plants. They convert inorganic minerals into the monobasic or dibasic ions.

The secretions of mobilizing Bacteria like citric acid, lactic acid and other organic acids helps in the mobility of the minerals towards the plant roots.

Specific functions of mobilizing bacteria

They perform their specific functions in three forms:

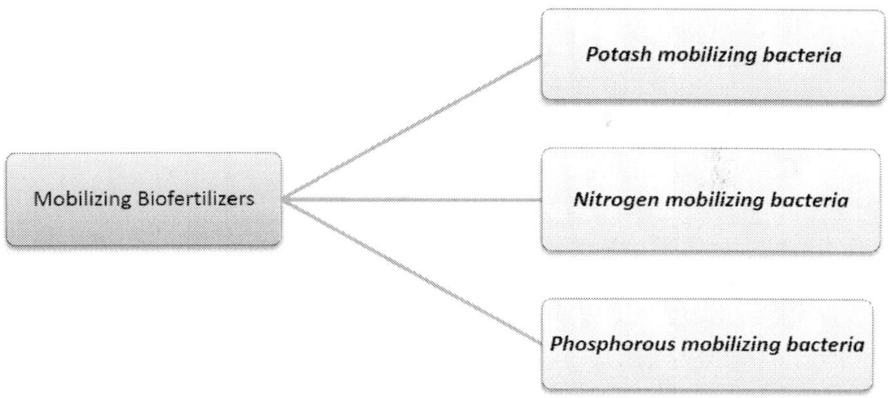

Figure: Different forms of mobilizing bacteria

i. Nitrogen mobilizing bacteria

   a. They help in fast growth of plants. The leguminous plants especially require high quantity of nitrogen.

ii. Potash mobilizing bacteria

   b. They play an important role in forming the protein/ amino acids from the ammonium ions present in the soil. They are also important for the transfer of proteins and carbohydrates etc. It regulates the cell membrane's permeability. It helps in activating number of enzymes like alcohol. These bacteria increase the resistance in plants to dry and hot weather without it the plat withers and become dry.

b. The potassium deficiency decreases the photosynthesis activity. These bacteria are in abundant number in soil.

iii. Phosphorous mobilizing bacteria

c. They convert insoluble phosphate HPO_4^{2-} to soluble phosphate $H_2PO_4^{-}$. These are least available in soil.

Microbes involve in mobilizing biofertilizers

The mobilizing biofertilizers are mainly present at the rhizospheres of the leguminous plants.

They mainly include

i. ***Nitrogen mobilizing bacteria***

a. The Bacteria used are Rhizobium and Azosprillum which are nitrogen mobilizing bacteria. They help in fast growth of plants.

ii. ***Potash mobilizing bacteria***

a. They are 3000 to 10,000 kg/ha of potash mobilizing bacteria in the upper 0.2 meter of soils. It includes *Bacillus Licheniformis* bacteria.

iii. ***Phosphorous mobilizing bacteria***

a. Agrobacterium, Trichoderm, Enterobacter, Aspergillus, Pseudomonas, Glomus and *Bacillus* are commonly involved in the mobilizing of the phosphorous from the soil to the roots.

Advantages

- ***Increased withstanding capacity***

The mobilizing biofertilizers increase the plant's capacity to withstand, droughts, cold, hot scorching summer heat, pests and diseases due the fact that the bacteria make the plants healthy.

- ***Toxins free***

  The mobilizing biofertilizers are toxins free which make them beneficial for the plants and environment. They do not contribute in the formation of the acid rain, ozone depletion and ground water contamination caused by the release of NOx in the environment. From the conventional fertilizers.

- ***Increase in productivity***

  The mobilizing biofertilizers increases the sucrose content in the plant, which is beneficial especially in sugar industry. They increase the crop yield by 20 to 30 percent and also increase the shelf life of the crop and its coloration.

Phosphate Solubilizing Microorganisms:

Phosphate solubilizing microorganisms are helpful microorganisms with the ability to solubilize inorganic phosphorus from variety of insoluble compounds. P-solubilization ability of rhizosphere microorganisms is measured to be among the most significant traits which associated with plant phosphate nutrition. PSB have been introduced to agro-commodities as phosphate biofertilizer. Phosphorus (P) is one of the main essential macronutrients for plants and is applied to soil in the form of phosphate fertilizers. However, a large portion of soluble inorganic phosphate which is applied to the soil as chemical fertilizer is immobilized quickly and thus not available to plants. In present ear, the major focus of concern in managing soil phosphorus is to optimize agricultural production as well as to minimize P loss from soils. PSB have attracted the attention of agriculturists as soil inoculums to improve the plant growth and yield. Phosphorus is among most important elements in plant nutrition. Adverse environmental effects of chemical based P fertilsers, depleting resources of high grade Phosphatic rocks and their skyrocketing prices have forced scientists to research for a sustainable and ecofriendly approach for efficient P availability in agriculture in order to fulfill the ever increasing demand of food. Soil microorganisms are involved in a range of processes which affect P transformation and thus influence the subsequent availability of P (as phosphate) to plant roots.

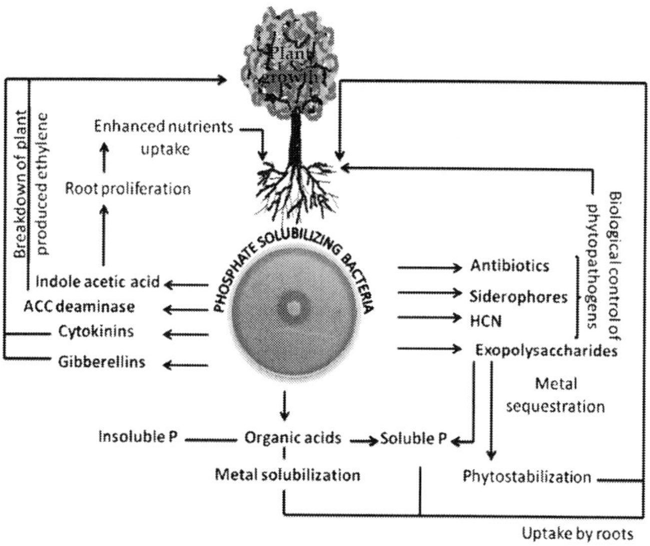

Figure: Role of Phosphate Solubilizing Microorganisms (Source: Almas Zaidi, Mohammad Saghir Khan, Munees Ahemad, Mohd Oves, and P.A. Wani, Microbial Strategies for Crop Improvement, Recent Advances in Plant Growth Promotion by Phosphate-Solubilizing Microbes, chapter 2, Springer-Verlag Berlin Heidelberg 2009, p 31)

Particularly, microorganisms can solubilize as well as mineralize P from inorganic and organic pools of total soil. The effective and smart use of PSM (phosphate-solubilizing microorganisms), opens up several new horizon for better crop productivity in addition to sustaining soil health. Though, the viability and sustainability of PSM technology is greatly dependent on the production and distribution of excellent quality inoculants to agro communities. Thus, there is an urgent need for broad and reliable research efforts in order to identify as well as to characterize more PSM with improved effectiveness for their eventual application under field conditions. Soil scientists as well as microbiologists bear's high responsibility towards society to find novel ways and means as to how soil as well as P could be enhanced without real application of chemical P fertilizers under different agro-climatic regions of the world. The promise of exploiting soil microorganisms to increase mobilization of soil P is still there. This target will be achieved only through better management of soil microbial communities, by development of more effective microbial inoculants, through the genetic manipulation of specific organisms, or with a combination of these approaches is not known. Soil microorganisms play an important role in the mobilization of soil P and that detailed understanding of their contribution to the cycling of P in soil-plant systems is required for the development of sustainable agriculture and our movement from a green revolution to an evergreen revolution can be accomplished. Phosphorus is the element which is abundantly found in both organic as well as inorganic forms in the soil. Phosphorus is among prime limiting factors for plant growth. The approaches of soil inorganic phosphorus in the rhizosphere vary significantly with plant species, status of nutrition's in targeted soil as well as different ambient soil conditions. Phosphate-solubilizing microorganisms (PSM) are capable of playing vital role in order to supply phosphate to plants in a diverse and sustainable was just to avoid phosphorus deficiency,. The solubilization of phosphate compounds by naturally abundant PSM is quite frequent under in vitro conditions. The variability in the performance has thus greatly hampered the large-scale application of PSM in sustainable agriculture.

AM fungi

An arbuscular mycorrhizal fungus (AM fungi) is a common type of mycorrhiza in which the fungus penetrates the cortical cells of the roots of a vascular plant. Arbuscular mycorrhizas are mosly characterized due to the formation of unique structures, arbuscules and vesicles by fungi of the phylum Glomeromycota. AM fungi assess plants to detain nutrients such as phosphorus, sulfur, nitrogen and micronutrients from the soil. Studied proved that the development of the arbuscular mycorrhizal symbiosis played a crucial role in the initial colonization of land by plants and in the evolution of the vascular plants. This symbiosis is a greatly evolved mutualistic relationship present between fungi and plants, the most prevalent plant symbiosis known. An arbuscular mycorrhizal fungus is present in 80% of vascular plant families. The marvelous advances in research on mycorrhizal physiology as well as ecology since last 40 years have led to a superior understanding of the multiple roles of AMF in the ecosystem. This information is applicable to human endeavors of ecosystem management, ecosystem restoration as well as of agriculture.

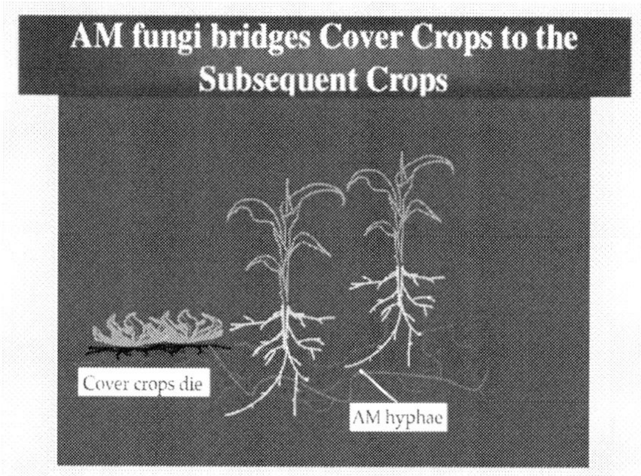

Figure: AM Fungi Role (Source: Z. Kabir, Ph.D. Soil Health Specialist for CA, NV & PIA USDA-NRCS, Davis, CA Arbuscular Mycorrhizal Fungi (AMF) in Soil Health Paradigm, 71st SWCS International Annual Conference July 24-27, 2016)

The nature of the relationship between plants and the ancestors of arbuscular mycorrhizal fungi is contentious. There are two hypotheses: (i) Mycorrhizal symbiosis evolved from a parasitic interaction that developed into a mutually beneficial relationship (ii) the second hypothesis is Mycorrhizal fungi developed from saprobic fungi that became endosymbiotic. Both saprotrophs and biotrophs were found in the Rhynie Chert but there are only few evidence to support either hypothesis. Few fossil evidence suggested that the parasitic fungi did not kill the host cells immediately upon invasion, although a response to the invasion was observed in the host cells. This response may have evolved into the chemical signaling processes required for symbiosis. In either case, the symbiotic plant-fungi interaction is considered to get evolved from a relationship in which the fungi was taking nutrients from the plant into a symbiotic relationship where the plant and fungi exchange nutrients.

AM fungi are obligate symbionts with limited saprobic ability and they largely depend on the plant for their carbon nutrition. AM fungi take up the products of the plant host's photosynthesis as hexoses. Transfer of the carbon content from plant to fungi may occur through the arbuscules or intraradical hyphae. Secondary synthesis from the hexoses by AM occurs in the intraradical mycelium. Inside the mycelium, hexose is converted to trehalose and glycogen. Trehalose and glycogen are a carbon storage form that can be speedily synthesized and degraded and may buffer the intracellular sugar concentrations. The intraradical hexose enters the oxidative pentose phosphate pathway, which produces pentose for nucleic acids. Lipid biosynthesis also occurs in the intraradical mycelium. Lipids are then stored or exported to extra radical hyphae where they may be stored or metabolized. The breakdown of lipids into hexoses, known as gluconeogenesis, occurs in the extraradical mycelium. Approximately 25% of the carbon translocated from the plant to the fungi is stored in the extraradical hyphae. Up to 20% of the host plant's carbon may be transferred to the AM fungi. This thing represents the host plant's considerable carbon investment in mycorrhizal network and contribution to the below-ground organic carbon pool.

Silicate Solubilizing Bacteria (SSB)

Microorganisms are able to degrade silicates. The metabolism of microbes numerous organic acids are produced and those have a twin position in silicate weathering. They supply H^+ ions to the medium and promote hydrolysis and the natural acids like citric, oxalic acid, Keto acids, and hydroxyl carbolic acids which form complexes with cations, promote their elimination and retention within the medium in a dissolved state. Silicon a naturally imperative component, required in living beings in a less sum. To the plants it gives regular barrier against numerous contagious illnesses, thus silicon accessibility to the plants is especially critical. Silicon is second most inexhaustible component found on the earth, but its solvency in soil is a noteworthy concern. In this way ID of proficient bacterial strains which could solubilize silicate is of fundamental significance. In addition, solublization of silica additionally discharges a few other basic supplements in the dirt. The component by which silicate solubilizing microorganisms (SSB) smothers different contagious maladies in plants is added to the resistance given by these microbes to the pathogens in entering host tissue. The term is characterized as Induced Systematic Resistance (ISR) and is powerful in stifling different rice phytopathogens including impact ailment. Expanding the proficiency of such strains in prompting resistance is a range to center and execute. Different reviews have been done including the increase of the Induced Systemic Resistance (ISR) of rice plant against Pyricularia oryzae, the most ruinous phytopathogen. In this review, they segregated ten strains specifically SSB-1 to SSB-10 set apart by their capacity in silicate solubilization and adversarial action against P.oryzae.In other review distributed in Canadian Journal of Microbiology, a SSB strain Q-12 was disengaged from the surface of feldspar and recognized as Bacillus globisporus Q12 in view of the 16S rDNA quality arrangement examination. They were likewise checked for their potassium solubilizing movement which has given positive reaction, the action was shockingly better on biotite than on feldspar and muscovite. There have been studies on bacterial populace in Loch Leven water. These microorganisms were found to upgrade the solubilization of silica from non-axenic societies of Asterionella Formosa, Tabellaria flocculosa, Navicular pellicu-losa and C. meneghiniana, contrasted and control societies cleaned with mercuric chloride.

COMPOST BIOFERTILIZERS

Main function

Compost fertilizers are those biofertilizers that make the best use of organic waste present in animal manure and food waste to make soil full of nutrients. They simply breakdown the complex organic compounds into simpler structure which plants can easily take up through roots.

Figure: Compost as biofertilizer (source: https://www.linkedin.com/pulse/organic-farming-techniques-dr-tohid-nooralvandi)

Microbes involved in compost biofertilizers

The micro-organisms like *Abcteria* and other *saprophytic bacteria* helps in breaking dead organic matter. The *Azetobactor* cultures are used in compost biofertilizers.

Also, the fungi are also responsible for breaking down the cellouse and lignin present in animal dung which is left undigested by the animals. The *Cellulytic fungal culture* is used for compost biofertilizers.

Ways of forming compost biofertilizers

- ***Animal manure***

  Animal manure is rich in the Nitrogen and other nutrients, which is used since centuries to grow crops in developing countries. They are used as the soil conditioner but provide nutrients slowly, because it needs to get break down first into smaller soluble compounds to be taken up by the plants.

- ***Urine***

  Urine contains high amount of urea which is a natural fertilizer for plants and is easily taken up by the plant roots in soluble forms. Urine is used along with other semi solid or solid contents to be applied onto the fields.

- ***Sludge***

  The sludge from water treatment plants can be utilized for the compost because it contains many nutrients essential for plant growth.

- ***Plant based waste***

  The plant based food waste basically contains the nutrients captured by the plants which on dissolution by the microorganisms releases many nutrients into the soil.

- ***Sewage water***

  Just like urine the sewage water is also utilized with other components of compost. The sewage water has high content of phosphorous and nitrogen present in it.

- ***Feathers***

The keratin present in the bird's feathers is a protein which can used as a compost.

Figure: Final Compost of feathers

Method and material

To form compost biofertilizer the materials used are allowed to be exposed to the good quantity of air/oxygen on daily basis and are placed in the containers with polythene bags covering the openings to prevent contamination. They are kept for 15-30 days for degradation at various sets of temperature and pH. The moisture content varies from 30 to 70 %.

Advantages

- ***Enhanced soil activity***

The compost biofertilizer have large quantity of beneficial bacteria, when applied to the crops helps in increasing the soil activity and hence help the plant in obtaining nutrients.

- ***Retain water***

The compost biofertilizer helps in retaining the water in soil also it also contains moisture which is helpful in the soil microbes to grow and make soil healthy. It is also helpful for the plants to take in water.

- ***Feeds the microbes***

The compost biofertilizer feeds the soil food web by providing enough available nutrients to the microorganisms and plants.

- ***Soil treatment***

  This biofertilizer treats the soil and replenishes its nutrients.

Plant Growth promoting Rhizobacteia:

Plant development advancing rhizobacteria (PGPR) were first characterized by Kloepper and Schroth to depict soil microbes that colonize the foundations of plants taking after immunization onto seed and that upgrade plant growth. The accompanying are certain in the colonization procedure: capacity to survive vaccination onto seed, to duplicate in the sperm sphere (locale encompassing the seed) in light of seed exudates, to join to the root surface, and to colonize the creating root system. The incapability of PGPR in the field has regularly been credited to their powerlessness to colonize plant roots. An assortment of bacterial qualities and particular qualities add to this procedure, however just a couple have been recognized. These incorporate motility, chemotaxis to seed and root exudates, creation of pili or fimbriae, generation of particular cell surface segments, capacity to utilize particular segments of root exudates, protein emission, and majority detecting. The era of mutants modified in articulation of these characteristics is supporting our comprehension of the exact part everyone plays in the colonization process. Advance in the distinguishing proof of new, beforehand uncharacterized qualities is being made utilizing nonbiased screening procedures that depend on quality combination advances. These procedures utilize correspondent transposons and in vitro expression innovation (IVET) to recognize qualities communicated amid colonization. Utilizing sub-atomic markers, for example, green fluorescent protein or fluorescent antibodies, it is conceivable to screen the area of individual rhizobacteria on the root utilizing confocal laser filtering microscopy. This approach has additionally been joined with a rRNA-focusing on test to screen the metabolic action of a rhizobacterial strain in the rhizosphere and demonstrated that microorganisms situated at the root tip were generally active.PGPRs improve plant development by immediate and backhanded means, however the particular systems included have not all been well-characterized, Direct components of plant development advancement by PGPRs can be shown without plant pathogens or other rhizosphere microorganisms, while roundabout instruments include the capacity of PGPRs to diminish the destructive impacts of plant pathogens on product yield. PGPRs have been accounted for to specifically improve plant development by an assortment of components: obsession of barometrical nitrogen exchanged to the plant, creation of siderophores that chelate iron and make it accessible to the plant root, solubilization of minerals, for example, phosphorus, and combination of phytohormones. Coordinate improvement of mineral take-up because of increments in particular particle fluxes at the root surface within the sight of PGPRs has likewise been accounted for. PGPR strains may utilize at least one of these systems in the rhizosphere. Sub-atomic methodologies utilizing microbial and plant mutants adjusted in their capacity to incorporate or react to particular phytohormones have expanded comprehension of the part of phytohormone union ass an immediate instrument of plant development upgrade by PGPRs. PGPR that blend auxins and cytokinins or that meddle with plant ethylene amalgamation have been distinguished.

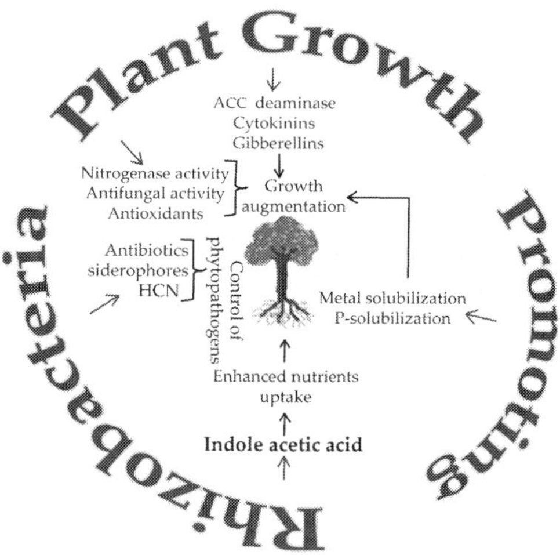

Figure: Plant Growth promoting Rhizobacteia-functioning (Source: https://perticos.wordpress.com/2016/02/10/plant-growth-promoting-rhizobacterial-forms/)

Liquid Biofertilizers

Liquid biofertilizers is progressively accessible in the market as one of the other options to synthetic and natural composts and in addition strong substrate-based biofertilizers. One of the advantages from biofertilizer is the commitment from populace of microorganisms accessible. These microorganisms may upgrade the plant development and make sound rhizosphere. The benefit of a fluid biofertilizer is that no strong bearer is required. These items are additionally created for potential application in modem horticulture, for example, soilless cultivating frameworks, viz. fertigation and hydroponics. Customarily, fluid bio manure is delivered from maturation of powerful microorganisms which was prescribed to be utilized inside three months. In this way, the advancement of minimal effort and long timeframe of realistic usability fluid biofertilizers was led at Malaysian Nuclear Agency (Nuclear Malaysia). Three biofertilizer inoculums (phosphate solubilizing microscopic organisms and plant development advancing microorganisms) were produced into four details of fluid biofertilizers. The fluid biofertilizers were kept at low temperatures (9 + 2 °C) and room temperatures (28 ± 2°C) for timeframe of realistic usability ponder. Supplement juices fluid biofertilizer kept at low temperatures demonstrated essentially high survival rates after capacity for six months when contrasted with different details and medicines. By and by, biofertilizers are set up as transporter based plans and lignite is the most generally utilized bearer material. The despicable cleansing of transporter material and taking care of while blending the microscopic organisms with bearer and pressing fill in as wellspring of defilements. Along these lines, the inoculant pocket could not hold desiredbiofertilizer life form for long time, which lessened the time span of usability of the biofertilizers. Further, the nature of the biofertilizer gets crumbled. In view of these reasons, biofertilizer application couldn›t ready to give practical outcomes in the fields. To keep away from these and to expand the quality and timeframe of realistic usability of bio inoculants, fluid detailing of Azospirillum and phosphobacteria was created. Additives were included with the bacterial societies and stuffed in quality PP bottles. The timeframe of realistic usability of fluid inoculants year, which can hold the cell tally of more than 108 for every ml with no contaminants.

Importance

In the bearer based (strong) bio-composts, the microorganisms have a time span of usability of just six months. They are not tolerant to UV beams and temperatures more than 30 degrees. The populace thickness of these organisms is just 108 (10 crores) c.f.u/ml at the season of creation. This tally decreases step by step. In the fourth month it decreases to 106 (10 lakhs) c.f.u/ml and toward the finish of 6 months the check is nearly nil. That is the reason the transporter based bio-composts were not viable and did not wind up plainly prevalent among the farmers. These deformities are amended and satisfied on account of Liquid Bio-manures. The time span of usability of the microorganisms in these fluid bio-manures is two years. They are tolerant to high temperatures (55 degrees) and ultra violet radiations. The consider is high as 109 c.f.u/ml, which is kept up steady up to two years. In this way, the use of 1 ml of fluid bio-manures is proportionate to the use of 1 Kg of 5 months old bearer based Bio-composts (1000 times). Since these are fluid definitions the application in the field is likewise extremely straightforward and simple. They are connected utilizing hand sprayers, control sprayers, fertigation tanks and as basal excrement blended alongside FYM and so forth.

Advantages of Liquid Bio Fertilizers:

The fluid inoculants have higher quality, longer time span of usability without contaminants than bearer based inoculants. The execution of fluid inoculants in the field is keeping pace with regular bearer based inoculants. Different favorable advantages of Liquid Bio-composts:

- Contains uncommon cell protectants or substances that support development of resting spores or blisters.
- Contains uncommon supplements that guarantee longer time span of usability, Better survival on seeds and soil and resistance to unfavorable conditions.
- Fluid arrangement guarantees that the item is anything but difficult to deal with and apply.
- Since the living beings are balanced out amid generation, dispersion and capacity, the action is improved after the contact and communication with the objective harvests.
- No loss of properties because of capacity up to 55 degrees.
- More prominent possibilities to battle with local populace.
- Doses are 10 times lesser than the bearer based bio-composts.
- High enzymatic action since tainting is nil.

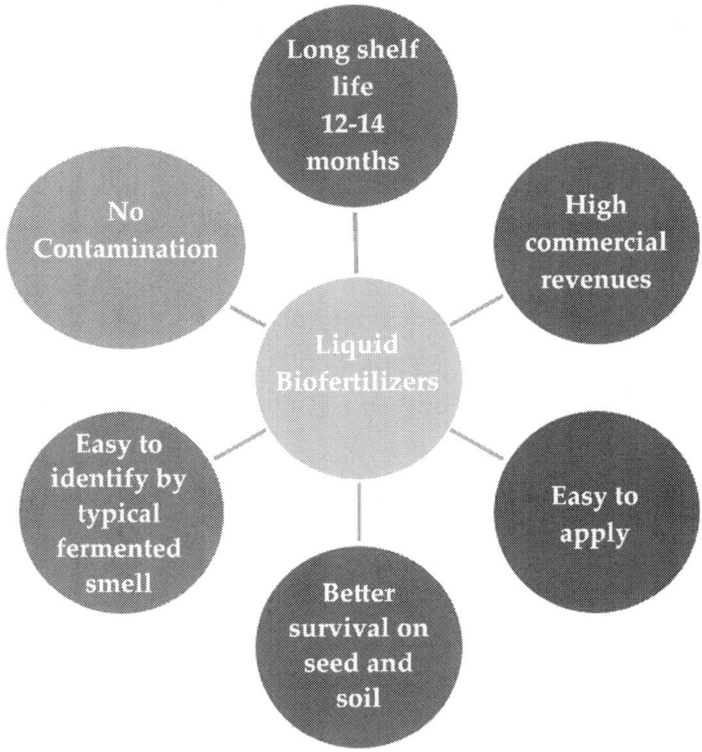

Figure: Other common advantages

Applications of Liquid Bio filters:

These are the application of liquid bio filters

Seed treatment

Seed treatment 250 ml each Azospirillum and Phosphobacteria per ha seeds; Seedling plunge 500 ml each Azospirillum and Phosphobacteria per ha seedlings; Soil application - 500 ml each Azospirillum and Phosphobacteria per ha seedlings; Tree crops 100 ml each Azospirillum and Phosphobacteria per tree once in six months.

Drinking water

For drinking water, organic water treatment includes the utilization of actually happening microorganisms in the surface water to enhance water quality. Under ideal conditions, including moderately low turbidity and high oxygen content, the creatures separate material in the water and along these lines enhance water quality. Moderate sand channels or carbon channels are utilized to give a support on which these microorganisms develop. These natural treatment frameworks successfully lessen water-borne infections, broke down natural carbon, turbidity and shading in surface water, in this way enhancing general water quality.

Wastewater

Bio filtration is utilized to treat wastewater from an extensive variety of sources, with changing natural arrangements and fixations. Numerous cases of bio filtration applications are portrayed in the writing. Bespoke bio filters have been produced and marketed for the treatment of creature wastes, landfill leachates, dairy wastewater, and residential wastewater. This procedure is flexible as it can be adjusted to little streams (< 1 m_3/d, for example, on location sewage and also to streams created by a district ($> 240\,000$ m_3/d). For decentralized residential wastewater generation, for example, for detached homes, it has been shown that there are imperative day by day, week by week and yearly changes of water powered and natural generation rates identified with current families' lifestyle. In this unique situation, a bio filter situated after a septic tank constitutes a vigorous procedure ready to maintain the fluctuation seen without trading off the treatment execution.

Use in aquaculture

The utilization of bio filters is normal in shut aquaculture frameworks, for example, recycling aquaculture frameworks (RAS). Many plans are utilized, with various advantages and downsides; however the capacity is the same: diminishing water trades by changing over smelling salts to nitrate. Smelling salts (NH_4^+ and NH_3) start from the brachial discharge from the gills of oceanic creatures and from the disintegration of natural matter. As smelling salts N is very harmful, this is changed over to a less poisonous type of nitrite and afterward to an even less dangerous type of nitrate. This "nitrification" prepare requires oxygen (vigorous conditions), without which the bio filter can crash. Additionally, as nitrification cycle produces H^+, the pH can be lowered which requires the utilization of cushions, for example, lime.

Bio fertilizers for micronutrients

Bio fertilizers are catagorized into various types depending upon their nature as well as function of fertilizer. Bio fertilizers for micro nutrients have zinc and silicate solubilizers in it which are given here.

Silicate solubilizing bacteria

Microorganisms are fully able to degrade silicates as well as aluminum silicates. During metabolism of these microbes various organic acids are formulated and they have dual role in silicate weathering. They supply H^+ ions to the medium and also accelerate hydrolysis. The organic acids for example citric acid, oxalic acid, Keto acids and hydroxy carbolic acids who from complexes with H^+, encourage their removal and retention in the medium in a dissolved state. Various researches performed on a Bacillus sp. isolated from the soil of granite crusher yard showed that the bacterium is able to dissolve several silicate minerals under *in vitro* condition. The examination of anthrpogenic materials like cement, agro inputs like super phosphate and rock phosphate exhibited silicate solubilizing bacteria to a varying degree. The bacterial isolates made from different locations had different ratio of silicate solubilizing potential. Soil inoculation studies with selected isolate with red soil, clay soil, sand and hilly soil showed that the organisms multiplied in about all types of soil and released more of silica and the accessible silica is increased in soil and water.

Zinc solubilizers

The nitrogen fixers, for example, Rhizobium, Azospirillum, Azotobacter, BGA and Phosphate solubilizing bacteria like B. magaterium, Pseudomonas striata, and phosphate mobilizing Mycorrhiza are known to be potential bio-fertilizers. However they are capable to supply only major nutrients but a host of microorganism that can transform micronutrients are there in soil which may be utilized as bio-fertilizers to supply micronutrients like zinc, iron, copper etc. The zinc can be solubilized by microorganisms viz., B. subtilis, Thiobacillus thioxidans and Saccharomyces sp. These microorganisms can be used as bio-fertilizers for solubilization of fixed micronutrients like zinc. From the finding of some very important studies it is deduced that a Bacillus sp.

(Zn solubilizing bacteria) can be used as bio-fertilizer for zinc or in soils where native zinc is higher or in conjunction with insoluble cheaper zinc compounds like zinc oxide (ZnO), zinc carbonate (ZnCO3) and zinc sulphide (ZnS) instead of costly zinc sulphate.

Plant growth promoting rhizobacteria

The rhizosphere is the narrow zone in the soil specifically influenced by the root system. This zone is rich in the protein nutrients and carbohydrates sugar necessary for the reproduction of bacteria due to the accumulation of variety of plant exudates. The rhizosphere is populated with the diverse range of microorganisms and bacteria colonizing this habitat are known as rhizobacteria. Plant living bacteria can be classified into beneficial, deleterious and neutral depending upon the effect of bacteria on plant growth and development. Beneficial free living soil bacteria are usually referred as plant growth promoting rhizobacteria. Plant growth promoting rhizobacteria affect plant growth in two ways either directly or indirectly. In direct plant growth promotion, PGPR either supply plant with compound known as phyto harmone that is synthesized by the bacterium or they facilitate the upgrade of certain nutrients by the plants from soil.

In indirect plant growth promotion, bacterium in this zone lessen or prevent the deleterious effect of phytopathogenic microorganism on plant by producing substance known as antagonistic substances or by inducing resistance to pathogens. A particular PGPR may affect plant growth and development by using any one, or more, of these mechanisms.

PGPR and their interactions with plants are exploited commercially and hold great promise for sustainable agriculture. Applications of these associations have been investigated in maize, wheat, oat, barley, peas, canola, soy, potatoes, tomatoes, lentils, radicchio and cucumber. Recent advance in the science of PGPR is at the stage where genetically modified PGPR can be produced. PGPR with antibiotic, phytohormone and siderophore production can be made.

Advantages of PGPR

- PGPR to control wide range of phytopathogenic fungi, bacteria, viruses, nematodes etc. Bio control mechanism which may be due to antibiotics, siderophores and enzymes or by systematic resistance against phytopathogen throughout the entire plant system.

- Effective against root rot and stem rot caused by Scelerotinia, Rhizoctonia, Phythium, Phytopthora, Alternaria, Cercospora, Macrophomina diseases etc..
- Improving plant stress tolerance to drought, salinity and alkalinity condition.
- Production of phytohormones like IAA, GA and Gibberlins
- It improves height and color for more production.

3 CHAPTER

FREE LIVING NITROGEN FIXING BACTERIA

INTRODUCTION

The crop productivity is increased by increased biological nitrogen fixation, increased plant growth due to nutrients adsorption and stimulation, the nutrientsup taking ability is also enhanced because of the reason that Nitrogen bio-fertilizers are used which induce hormones or anti-biosis or that they might decompose organic residues. The extensive application of chemical fertilizers is prevented by using nitrogenous bio-fertilizers which reduces the cost and amount of chemical fertilizers and results in preventing the environment from pollution. Sustainability of farms can be achieved by using biological and organic fertilizers thus keeping the input system low.

Free Living Nitrogen Fixing Bacteria

The bio-fertilizers convert nutritionally important components like nitrogen, phosphorous from unavailable to available form through nitrogen fixation and solubilization of rock phosphate. Because bio-fertlizers are living cells of different types of microorganism that colonize the rhizosphere and promotes plant growth when applied to seed and plant surface or to soil. The main reserves of nitrogen is atmospheric nitrogen which is found in molecular form, due to this factor fixed nitrogen is a limiting nutrient in most environment. The biofertlizers are potential biological keys to sustainable productivity of soil and cost effect for farmers as well as eco-friendly by protecting the environment.

The prokaryotic cells biologically fix nitrogen through nitrogen fixation process which makes easy for plants to assimilate the molecular nitrogen. For many years, there were very few bacteria known to degrade fix nitrogen from the environment but in last 30 years, the nirtrogen fixation is a representative property of bacterium phyla and also in Methanogenic Archaea.

Phylogeneticall Bacteria Types

There are two groups of bacteria that are phylogenetically related and they symbiotically fix nitrogen within nodules of vascular plant.

The two groups are:

i. **Rhizobia** (Alpha-Proeobacteria) this group of bacteria is associated with leguminous plants belonging to the super family of angiosperm (Fabaceae).

ii. **Frankia** (In-Actinobacteria) which includes wide range of plants from eight separate plant families.

iii. **Cynobacteria** is a very important group of nitrogen fixing bacteria which is found in association with higher and lower plants as well as it deals with algae and fungi.

The colonization of root surface of non-leguminous plants, without differentiated structures are diverse range of associative symbiotic nitrogen fixing species. There is a new category of nitrogen fixing endophytes due to the frequent isolation of bacteria from sterilized root surface.

The Nitrogen Fixation Process

The molecular nitrogen is converted into ammonia with the help of enzymetic action, catalyzed by Nitogenase enzyme. This is an oxygen labile enzyme complex which is highly conserved in symbiotic and free living diazotrophs. The most common nitrogenase is called Mo-

Nytrogenase or it can be referred to as conventional nitogenase because it contains the prosthetic group with molybdenum along with iron and cobalt groups attached. FeMoCo (Iron-Molybdenum-Cobalt group).

There are some bacteria like Azotobacter and some cyanobacteria along with several photosynthetic nitrogen fixers that carry additional forms of nitrogenase having the co-factor Vanadium (V-nitrogenase) or the nitrogenase may contain iron as a co-factor (Fe-nitrogenase). The nitrogenase enzyme that has been purified from different sources is composed of two metalloprotiens.

In which the component 1, designated as Mo-Fe protein, is a tetramer of 220,000 Da composed of two non-identical sub-units. Two Fe MoCo are bound to subunits of the MoFeprotein. However, there are two other prosthetic groups containing 4Fe- 4S clusters. 'P-clusters' are covalently bound to cysteine residues of MoFeprotein bridging and subunits. The third type of Fe-S group is linked to the Fe protein. Nitrogen reduction is a very complex mechanism not as yet fully explained.

Nitrogen-fixing Bio-fertilizers

Through biological nitrogen fixation, plants use elemental nitrogen only. The transformation of the inert N_2 into organic compounds happens by the nitrogen fixing bacteria. The reason that nitrogen-fixing bacteria are used as bio-fertilizers is that they are composed of groups of microorganisms and microbial inoculants which fix nitrogen from atmosphere.

There are these groups of nitrogen fixing organisms that are categorized here:

i. Free-living bacteria (Aztobacter and Azospirillium)

ii. Symbionts (Rhizobium, Azolla and Frankia) along with Blue Green Algea.

iii. Non-legumes species (*Achromobacter, Acetobacter, Azomonas, Beijerinckia, Bacillus, Clostridium, Enterobacter, , Desulfovibrio ,Corynebacterium, campylobacter, Herbaspirillum, Klebsiella, Lignobacter, Mycobacterium Alcaligenes, Arthrobacter, Rhodospirillum, Rhodopseudomonas, Xanthobacter, Erwinia, Derxia, Mycobacteriumand Methylosinus.* associated with N_2 fixing bacteria.

The members of Aztobacter and Azospirillium genera have been tested under field conditionsto increase the yield of cereals and legumes. From the rhizosphere of various cereals, the isolation of several species and genera of N_2 fixing bacteria are achieved.

Nitrogen fixing Bio-fertilizers Rhizobium

From the family of Rhizobiaceae Rhizobium belongs, which fix nitrogen with legumes only from 50 to 100 kg/h and is symbiotic is nature. The roots are colonized by Rhizobiumin specific legumeswhich act as ammonia production factories to form root nodules which are tumour like growths. Rhizobium can symbiotically work with legumes and with Parasponia which is a non-legume. The nodulation of legume host is done by six species of Rhizobium which is based upon the crosss-inoculation studies.

A "**cross-inoculation group**" is a group that is classified by a collection of leguminous species that develop nodules on any member of particular plant group. This is the reason that single inoculation group includes all species that are infected by an individual bacterial strain. Azospirillumis a nitrogen fixing fertilizer from the family Spirilaceae, which is associative and heterotrophic in nature. They can produce growth regulating substances as well as they have the ability to fix nitrogen of about 2-0-40 kg/ha.

A.lipoferum and A.brasilence are the two mainly used species of this genus worldwide that has numerous benefits of inoculation. There are other members of this specie also such as A.amazonense, A.brasilense, A.halopraeferens. With many plants, the Azospirillum form associative symbiosis especially with those plants that have the C4-dicarboxyliac path way of photosynthesis which is also called Hatch and Slack pathway. Because these plants grow and fix nitrogen on salts of malic and aspartic which are organic acids in nature. That's why they are recommended for crops like; maize, sugarcane, sorghum and pearl millet etc. The penetration of Azotobacter into the root tissue happens in substantial proportion and does not colonize the roots only butit also remains on the root surface. Any visible nodules are not produced by them on root tissues that live in complete harmony with plants.

Azotobacter

Azotobacterbelongs to the family named Azoto bacteriaceace, which isheterotrophic, aerobic and also free living in nature. The most commonly found species of this family is A. chroococcum which is found in arable soil but many of them are present in alkaline andneutralsoils. A. vinelandii,A. macrocytogenes A. beijerinckii and A. insignisare other found species. Due to the presence of antagonistic microorganisms and lack of organic matter in the soil the A. Azotobacter hardly exceed the number in growth of 104-105g.-

Blue Green Algae (Cyanobacteria) and Azolla

Insole acetic acid is produced by blue green algea, as well as the production of auxin and Gibberllic acid from eight different families that are phototropic in nature. They are found in abundance in paddy fields as they fix 20-30kg N/ha of rice fields. This is the reason they are called "paddy organisms". In large quantity Nitrogen is required by them because of low land price production. For low land rice, the soil N_2 and BNF are the sources for nitrogen. The organic nitrogen is mineralized in soil and BNF occur not only due to free living bacteria but also through associated rice plant bacteria that are the sources of 50-60% nitrogen for these plants.

To achieve food security by practicing sustainable agriculture, the requirement for fixed nitrogen should be increasingly met by BNF rather than by the application of industrial nitrogen fixation. There are specialized cells called heterocysts which are micro nodules in action for nitrogen synthesis and fixing. The most nitrogen fixing BGA consist heterocyst which are in the form of long chains of filaments as vegetative cells including specialized cells. There is an association found by BGA forms which can fix nitrogen with flowering plants, ferns, liverworts and fungi. But the most common associations of symbiosis can be found in between Anabaena azollae (BGA), Azolla and free floating aquatic ferns.

The usage of Azolla as bio-fertilizer in rice crops is because of its quick decomposition in soil and efficient availability of nitrogen to rice plants which are the important factors. Besides providing nitrogen, these biofertilizers or bio manures contribute in providing Mb, K, P, Zn, S and Fe in significant amount along with providing other micronutrients. Azolla can be applied as green manure by incorporating in the fields prior to rice planting. A. pinnata is the most common species occurring in India and can be propagated on commercial scale by vegetative means.

Nitrogen-fixing Endophytes

Because Azoarcus sp, Gluconaceto bacter and Herbaspirillum are the most studied genera, they are found in occurrence with nitrogen-fixing bacteria within plant tissues of a host plant which does not show diseases symptoms. Without causing damage, the Endophytes multiply and spread within plant tissues. In the initial stages of infection, surface colonization at the site of emergence of root hairs happens just like those with rhizospheric bacteria.

For tissue penetration in Azoarcus, type IV pili are essential for that process and hydrolytic enzymes, or endo-glucanases. After the sterilization of root system, the recovered quantity of bacteriacan reaches about 108 CFU per g of dry weight. Another ability they have is systemic spreading of bacteria in case of sugar cane infection with G.diazotrophicus and C4- gramineous plant Miscanthus sinensis infection caused by H. frisingense. In plant xylem vessels this systemic spreading of bacteria is found and also it is found in shoots.

Rhizobium-legume Symbiosis Host plants:

With root-nodule bacteria, many leguminous plant species enter into symbiotic relationship all together which is called rhizobia. The three main subfamilies (family Leguminosae) are distinguished such as,

i. Caesalpinoidae

ii. Mimosoidae

iii. Papilionidae.

There are few nodulating members in Caesalpinoidae. The members of Papilionidae. Mimosoidae subfamily are important agricultural crops mostly, this sub-family has recently got attention. The Parasponia sp. is the only one non-legume, nodulatethe woody plant by rhizobia and alsoutilizes nitrogen fixed by the bacteria.

There are large and important divisions which are named Proteobacteria that contains rhizobia of gram-negative nature. Standard procedures are applied by following taxonomic classification of rhizobia which is based on the phylogeny of housekeeping genes and whole-genome similarities. It is not always possible to differentiate nodule formers by their names the reason being that nodulation functions cannot be evolved until after bacterial housekeeping properties.

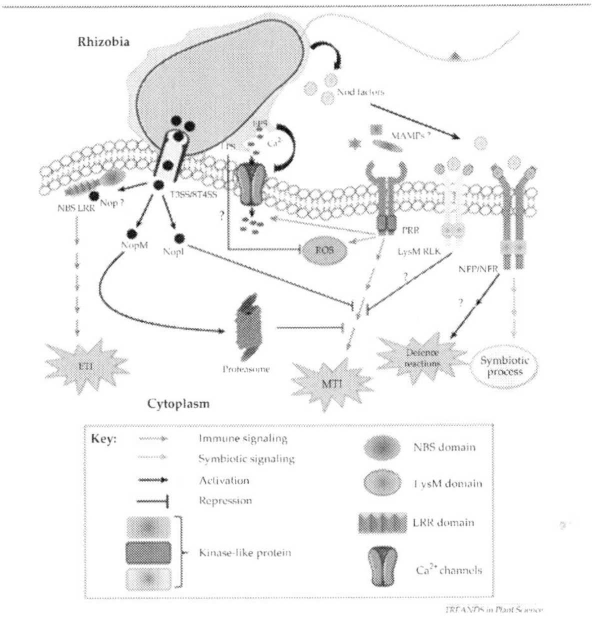

Figure: Rhizobium–legume symbioses (source: Benjamin Gourion, Fathi Berrabah, Pascal Ratet, Gary Stacey, Rhizobium–legume symbioses: the crucial role of plant immunity, Trends in Plant Science, 20 (3): 186-194, 2015)

Nod Factors and Nodulation Genes

There are nodulations of genes encoding and nitrogen fixation functions present on a mutual genetic determinant form rhizobia. These genes are carried on plasmids quite often or they are carried on other accessory elements, such as symbiotic islands and properties encoded by them can be easily lost or found. There is a production of a nodulation signal, the Nod factor, which is a lipo-chitooligosaccharide and the nod, nol and noe gene products are involved in this production. There is a specific bacterial receptor called NodD, which is a transcriptional activator of other nodulation genes. When flavonoids are produced by host plants and related secondary metabolites in the rhizosphere, then those genes are perceived by this specific bacterial receptor. There is codding on the core of the Nod factor molecule which molecule is encoded by canonical nodA, nodB and nodC whereas, for example, nodFE take part in poly-unsaturation of the fatty acyl group attached to the core molecule. In the case of Nod factors produced by *Azorhizobium caulinodans, the* nodulation genes encode enzymes that add a variety of substituents to the core. By the triggering of developmental program that leads to the construction of the root nodule and entry of rhizobia into the nodule, The Nod factor acts as an elicitor of root nodule formation by the plant. The Nod factor is an important host specificity determinant. Recently it is found that, the rhizobium does not possess canonical nod ABC genes but it uses other mechanisms for signaling to the plant. This Nod factor paradigm is challenged by the one who discovered that photosynthetic, stem- and root-nodulating rhizobia. A purine derivative can play a role in triggering nodule formation instead of the Nod factor. That Hypothesis was made after many experiments were performed. Then this hypothesis pointed to the complexity of the symbiotic system that showed bacteria which had employed diverse strategies to gain entrance into the roots.

The nodule Organogenesis & infection process

The rhizobial infection is responded by the presence of compatible rhizobia through deformation (shepherd's crooks, cauliflower structures, etc. prior to the infection process via plant root hairs. The plant perceives the rhizobial signal and starts the developmental program aimed because of the formation of symbiotic nitrogen-fixing nodules at deformation stage. There is a set of plant genes which initially called nodulins were specifically activated in response to nodulation factor perception. Calcium acts as a secondary messenger in the infection process because calcium spiking was observed in the early stages of root hair infection. Further process stages in the signalling cascade takes to induction of cortical cell division. The rhizobia enters through plant-derived infection thread which is a tube formed to facilitate rhizobial entry into the deeper layers thus a nodule meristem is formed within the root. Finally the rhizobia wrapped into a plant derived membrane called symbiosome membrane are delivered while the infection threads grow transcellularly.

Nodule physiology

The host tissues form a specialized tissue which maintains an environment for nitrogen fixation to occur, during nodule formation. There are specialized organelles which are like forms of bacteria thus called bacteroids which are engulfed in plant-derived membranes, forming symbiosomes in the nodule. The plant gives energy required for the reduction of di-nitrogen inside the nodule.

The dicarboxylic acids give the bacteroids carbon and energy through the symbiosome membrane whereas, the Photosynthate present in the form of "sucrose" is transported to the nodule. A high flux but at the same time low internal concentration of oxygen is achieved with the help of leghemoglobin for generation of energy through respiration. Through the symbiosome membranes, the ammonia production in the bacteroid has to be transported to the plant.

In addition to ammonia Alanine is transported. While In the transport mechanism, amino acid flux reverse back through the symbiosome membrane that has been involved. In the plant cytosol, Ammonia assimilates further into glutamine or asparagine. In the uninfected cells just near to the infected cells, there is further conversion into ureides in determinate nodules. This conversion does not happen in indeterminate nodules, and rest of the plant cells is normally infected. The study of symbiosome biochemistry is impaired by technical difficulties involved when intact but isolated symbiosomes are used.

Actinorhizalsymbiosis Host plants

There are about 200 species of Actinorhizal among which 24 genera in eight angiosperm families are represented by Actinorhizol plants. From the order Actinomycetales and family Frankiaceae, the genus Frankia consists ofhigh mol% G+C Grampositive genera. Frankia is a filamentous bacterium having the slow growth rate characteristic, can form hyphal colonies without an aerial mycelium. It has ability to distinguish two unique development structures that are seriously needed for its survival and these striking features it has arevesicles and spores. Spores are contained in multilocular sporangia which are the reproductive structures of Frankia, while Vesicles are the site for actinorhizal nitrogen fixation.

Because of the slow growth rate of filamentous hyphae, the development of molecular genetic tools in Frankia aredifficult to implement as compared to that in rhizobia. The genetic transformation, mutagenesis and functional complementation failed to provide conclusive results in most cases.

Infection process

There are two modes of infection of actinorhizal plants described by Frankia:

i. Intercellular root invasion

ii. Intracellular root hair infection

The mode of infection is dependent on the host plant, just like the situation encountered in rhizobia. Intracellular infection via root hairs e.g. of Casuarina, Myrica & Alnusstarts with root hair curling while signal exchange between Frankia and the host plant. After conducting investigations in different laboratories, the signalling molecule pathway is not identified. The Frankia molecule can induce root hair curling in host plants. This initial characterization of a Frankia molecule shows that it different from Nod factors in rhizobia and it is consistent with the absence of nod genes in Frankia genome. The results declare that the infection proceeds intracellularly in the root cortex, after the invagination of growing filaments of Frankia in the curled root hairs.

The encapsulation of cell wall deposit to Frankia hyphae happens consisting of xylans, cellulose and pectins of host origin. Near the invading root hair in the cortex, limited cell divisions take place at the same time. This cell division leads to the formation of pre-nodule which is a small external protuberance. Cell divisions are induced in the pericycle, which is located opposite to the protoxylem pole as the pre-nodule develops thus giving rise to another nodule primordium. The distantly induced primordium constitutes the nodule whereas the actinorhizal prenodules do not evolve into nodules. The actual function of the pre-nodule was investigated in *Casuarina glauca*.

The cortical cell divisions take to the formation of a unique nodule primordium which evolves into a mature nodule where the sequential diversity of pre-nodules and then nodules create a major difference from the situation in legumes. The pre-nodule can be remaining form of a common nodule ancestor for legumes and actinorhizal plants or it can be a parallel symbiotic organ of its own.

The pre-nodules show the same ability as nodules and hence can be taken as very simple symbiotic organs according to the study of symbiosis- related gene expression coupled with cellular modification i.e., cell wall lignification. In the intercellular root invasion process *e.g.* Discaria, Elaeagnus, Hyppophae and Ceanothus, the Pre-nodule does not form. In the electron-dense matrix which is secreted into the intercellular spaces, the Frankia hyphae penetrate between two nearby rhizoderm cells and progress apoplastically from cortical cells. A second level of compartmentalization can be described in *Casuarina glauca* nodules based on accumulation of flavans just now, which happens in uninfected cells in the endodermis and cortex.

Cyanobacteria and symbiosis:

Cyanobacteria are widely found everywhere in aquatic and terrestrial environments. Oxygenic photosynthesis performed by them which is the reason they are regarded as algae. In five Sections based on morphological criteria mostly, they are classified into the Bacterial domain. Cyanobacteria are the largest and the most diverse group of Gram-negative prokaryotes. Basically they belong to the genus Nostoc and Anabaena, and their associations with plants are fundamentally limited to heterocyst us cyanobacteria Nostocales while nitrogen fixation can be found both in unicellular and filamentous species. There exists a diverse variety of non-vascular lower plant thus belonging to bryophytes, which includes liverworts, hornworts, algae and fungi, to develop associations with cyanobacteria, as well as many marine eukaryotes.

Mechanisms for plant growth promotion Phytohormones

Plant growth regulators take part in the growth and development of cells, tissues, organs, and also in the entire plant. In very small amounts, these compounds are active in plants and their production is extremely regulated. Plants produce phytohormones and also produce numerous plantassociated bacteria which are both beneficial and harmful, producing one or more of these substances. *Azospirillum* is famous for its ability to excrete phytohormones like gibberellins cytokinins and auxins, Among the PGPR species. in morphological and physiological changes of the inoculated plant roots, Many studies suggest the involvement of indole- 3-acetic acid (IAA). It is significant that IAA synthesis by *Pantoeaagglomerans is induced* bacterial plant dependent response.

Of the more actively phytohormone-synthesizing strains, Differential behavior of the isolates in relation to the proximity to plant tissues could be linked to a great competitiveness. To determine this issue, Inoculation experiments of single, or mixtures of strains, previously isolated from different distances from roots, could help. if the rhizosphere gradient of plant exudates participates in determining a differential response in the bacterial synthesis and release of phytohormones would also be exciting to determine. The phytohormone-mediated roles of bacterial ephyphitic communities on plants are yet not clear. The future of biofertilizers based on hormone-producing bacteria seems very promising. Bacterial participation raises the phytohormone levels in plants, which is showed by large numbers of experiments. This can be via bacterial synthesis or through bacterial induction of plant hormone synthesis but both offer economic and ecological advantages.

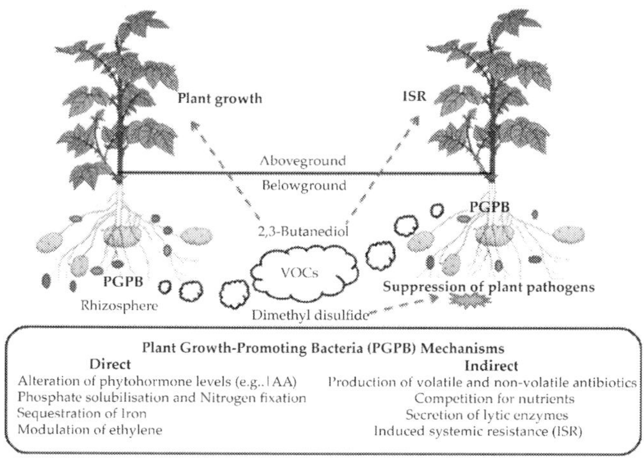

Figure: Mechanisms for plant growth promotion Phytohormones (source: Siva L. S. Velivelli & Angela Sessitsch & Barbara Doyle Prestwich, The Role of Microbial Inoculants in Integrated Crop Management Systems, Potato Research, 57:291–309, 2014)

ACC De-aminase activity:

Ethylene exposition induces reduction in the growth rate including different observable changes in plants. Since monocots are less sensitive to ethylene, this is especially true in stressed dicot plants. By lowering the plant ethylene levels, It has been proposed that PGPR may enhance plant growth. In these cases, 1- aminocyclopropane-1-carboxylate (ACC) is the immediate precursor of ethylene. This compound is hydrolyzed by bacteria-expressing ACC-deaminase activity. The ACC-degrading bacterium use products of this hydrolysis, Ammonia and ±- ketobutyrate as nitrogen and carbon sources.Bacteria belonging to phylogenetically distant genera such as Alcaligenes sp., Bacillus pumilus, Pseudomonassp. and Variovorax paradoxusas well as, Azoarcus, Azorhizobium caulinodans, Azospirillum spp., Gluconacetobacterdiazotrophicus, Herbaspirillum spp., Burkholderia vietnamiensis and others were identified by their ability to grow on minimal media containing ACC as sole nitrogen source.

Environmental Stress Relief:

Under restrictive environmental conditions, several associations between plants and beneficial bacteria show a protective response *e.g.* Wheat and fababeans when they are subjected to saline stress, they show greater growth when inoculated with *Azospirillum*, compared to non-inoculated plants. This favorable effect may be attributable directly to bacteria or indirectly to the effect on plant physiology.The production of microbial metabolites like polysaccharides has a positive effect on plants grown in water stress and modifies the soil structure. Growth parameters of sunflower plants under water stress inoculated with an exopolysaccharide (EPS)-producing Rhizobium sp. were greater than in uninoculated plants. After inoculation with an EPS-producing *Pantoea agglomerans* isolate, Promotion effect in wheat plants was also observed. The aggregation of rhizospheric soil depended on a bacterial polysaccharide that enlarged the amount of soil adhering to roots, in wheat plants inoculated with *Paenibacillus polymyxa.*To turn on particular metabolic activity, Bacteria can stimulate the plant by increasing its exudates, and consequently, improve rhizospheric soil qualities. Inoculation of *Arabidopsis* with *P. polymyxa* the water-stress gene ERD is switched on in the same way. In comparison to control plants, Inoculated plants show improved response against pathogenic colonization and drought stress. Hence, Inoculation induces protection against biotic agents, and also against abiotic ones. Through the bacterial release of soil structure-improving substances and by inducing the plant to activate stress responsive mechanisms, the PGPR can protect a plant, against aggressive environmental and particularly hostile soil conditions. The uses of bacteria that allow plants to thrive are probably the best option to obtain good yields at lesser ecological costs.

Enhancing phosphorus availability for plant growth

Phosphorus is a plant nutrient which is essential even in low availability for many agricultural soils. Many agricultural soils have high total Phosphorus today. The Phosphorus content is enhanced due to the application of P fertilizers over long periods of time. Contrary to that, most of the Phosphorus is found in mineral forms and is available slowly to the plants. Most of the insoluble Phosphorus is present as aluminum and iron phosphates in acid soils and calcium-phosphates in alkaline soils. The rhizosphere bacteria has the ability to solubilize insoluble Phosphorus minerals that has been attributed to their ability to reduce pH by the excretion of organic acids e.g. gluconate, citrate, lactate, succionate and reduce protons during the assimilation of NH_4^+. These microorganisms grow in tri-calcium phosphate mediaor any media similar to insoluble materials as the only phosphate source but also solubilize Phosphorus in abundant quantities for nutritional demands, thereby making it available for plants.

Siderophore production

Iron is a growth element essential for all living organisms. The low bioavailability of iron in soil habitats and on plant surfaces creates a dangerous competition. Under iron-limiting conditions PGPB produce "siderophores" which is a low-molecular-weight compound to acquire ferric ioncompetitively. Siderophore is a Greek word meaning "iron carrier" is small, it has high-affinity iron chelating compounds expelled by microorganisms such as bacteria, fungi and grasses.

To find iron from these mineral phases by formation of soluble Fe^{3+} complexes that can be taken up by active transport mechanisms, for this reason the siderophores are released by microbes. There are so many siderophores bio-synthesised independently, while many siderophores are non-ribosomal peptides. Siderophores are significant for some pathogenic bacteria also because of their acquisition of iron. With enterobactin being one of the strongestbinders to Fe^{3} from Siderophores to strongly bind Fe^{3+}. most of the isolates are from Gram- negative bacteria corresponding to the *Pseudomonas* and *Enterobacter* genera, and *Bacillus* and *Rhodococcus* genera are the Gram-positive bacteria found to produce siderophores exposed by the distribution of siderophore producing isolates according to amplified ribosomal DNA restriction analysis (ARDRA) groups.

4
CHAPTER

ENDOPHYTIC NITROGEN-FIXING BACTERIA AS BIO-FERTILIZER

INTRODUCTION

To feed the rapidly increasing population of the world, the methods of search and utilization of resources are adopted for endophytic bacteria and to understand their abilities to promote plant growth and productivity indeed shows the presence of natural associations of these having beneficial impact. The associative and endophytic bacteria failin large number under natural conditions without any gaps, regardless of the fact that they have shown potential in laboratory and green house conditions. When plants inoculated with endophytic bacteria were transported from pot to field conditions, there was noted reduction in the efficiency. The factors that affect colonization and the bacteria derived benefit to plants can be soil type, nutritional status of soil, host plant genotype and age as well as climatic conditions.

The plant growth and yield can be enhanced by the combined use of fertilizers and endophytic bacteria, evident from previous reports. This combined use of fertilizers can reduce the inputs of chemical fertilizers in the soil. Because of the regulatory mechanism acting in the diazotrophs, the high amount of available utilizable N results in reduced colonization of endophytic bacteria and also reduces the process of N_2fixation. Use of compost is useful to some extent because it provides utilizable N to support the growth of microorganism and make the plant evade from negative effects of diazotrophs colonization. The one of the major challenges is the selection of plant genotype, age and compatible associative bacteria. In order to boost the productivity by using specific bacterial strain, the understanding of this compatibility can be beneficial. Because the multiples field trials are needed to increase the parameters for getting the maximum output, the colonization of associative bacteria depends upon seasonal variations and soil hydric stress. Another factor which plays a critical role for plant defense mechanisms because it can limit or reduce the colonization of associative bacteria. The colonization is not a well understood mechanism. In order to generate the bacterial colonization process and to increase plant growth attributes and also to provide a better way to understand the ecology and attributes of the endophytic diazotrophs, intelligent analysis of genomic and functional genomic studies can be beneficial to manipulate the conditions.

Nitrogen fixing Bacteria as Bio-fertilizers:

Nitrogen is the most limiting nutritional factor for the growth of plants. Plants require exogenously fixed nitrogen for growth and development factors because plants cannot utilize atmospheric N_2as it is. To use nitrogen by plants, Atmospheric N_2 should be reduced to ammonia first. The chemical Nitrogen fertilizers can be used to provide nitrogen nutrition to plants practically. The contamination of underground and surface water happens due to leaching of nitrates and release of greenhouse gases when manufacturing in industries. These factors are associated with environmental hazards during the manufacture and use of Nitrogen fertilizers. Moreover, coal and petroleum are non-renewable resources required for the manufacturing of chemical fertilizers. For the improvement of the yield of commercial crops, the too much and prolonged use of chemical fertilizers can have negative effect on soil fertility and thus reduces their agricultural sustainability.

In a manner which is efficient and sustainable, there is a concern to necessitate the search for an alternative strategy to provide nitrogen nutrition to the plants. The biological nitrogen fixation is an alternate to chemical fertilizers because they have immense potential. The few members of the prokaryotic organisms fix biological nitrogen exclusively. Biological nitrogen fixation is a process where atmospheric N_2 is reduced to NH_3. The microbial enzyme nitrogenize catalyzes this process. The endophytic bacteria colonize the interior of plant without exerting any substantive harm to their host. The nitrogen fixing bacteria, has reference to endophytic bacteria while reviewing their application, properties, ecology, and advances in biology.

Nitrogen- fixing endophytic bacteria has to faces less competition when they can make available the fixed nitrogen directly to plants due to the reason that they are sheltered inside plant tissues as comparing to its rhizospheric counterparts. Moreover, for nitrogen fixation to be efficient, the partial pressure of oxygen inside the plant tissue is more yielding. Nitrogen fixing endophytic bacteria promotes plant growth by contributing up to 47% of nitrogen derived from air and these endophytic bacteriacan be isolated from different plant species. The total nitrogen difference method, acetylene reduction assay, analysis of nitrogen solutes in xylem and other plant parts and N-Labeling Methods are used to evaluate nitrogen fixing ability of bacteria. Also, molecular approaches e.g. amplification, analysis of nitrogen fixing genes (*nif*genes), and qualitative and quantitative estimation of the products.

Nitrogen Fixation by Endophytic Bacteria:

In last years to the various crop plants, application of endophytic bacterial inoculants supplying Nitrogen requirement has got lots of attention for increasing plant yield efficiently in a sustainable manner. In the interior habitats of plants, the efficient Nitrogen supply by endophytic diazotrophic bacteria in sugarcane and kallar grass suggests the possible avenues of biological nitrogen fixation. . In addition, bacteria isolated from non-leguminous plants fix the Nitrogen in endophytic manner like in rice, wheat, maize, sorghum plants.

The *Gluconoaceto bacterdiazotrophicus* (*Acetobacter diazotrophicus*) is the most important contributor of endophytic biological nitrogen fixation in sugarcane, and thus it has the ability to fix the Nitrogen approximately about 150 Kg N ha^{-1} $year^{-1}$, as an evident from the reports. *Azoarcus* is another popular potential N_2fixing obligate endophytic diazotroph. The *Azoarcus* increases the hay yield of about 20–40 t ha

$^{-1}$ year^{-1}without the addition of any Nitrogen fertilizer in saline sodic alkaline soils thus dwells in the roots of kallar grass.many energy plants (C 4 plants) which are best to support the Nitrogen nutrition of the plant and thus they can be found to harbour bacterial population. The *Herbaspirillum* sp., inoculated into rice seedlings maintained in Nitrogen free Hoagland solution containing 15 N-labelled N in a study, it showed 15 N dilution amounting upto 40% increase in total N of plant. Seed inoculation with a strain of *Rhizobium leguminosarum*bv*trifolii* in pot experiment stimulatesthe growth of wheat, corn, radish, mustard and some varieties of rice shoots. According to the investigations, endophytic diazotrophs have a considerable potential to increase the productivity of important cash crop plants and non-legumes.

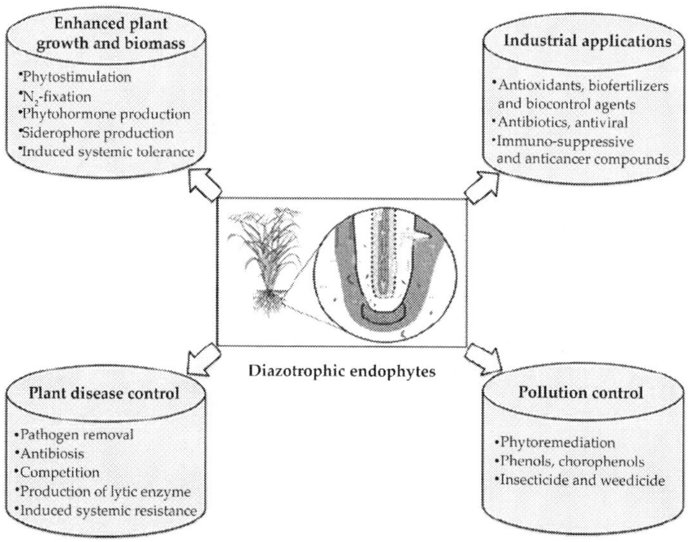

Figure: Applications of diazotrophic endophytic bacteria in different fields. (source: Garima Gupta, Jitendra Panwar, Mohd Sayeed Akhtar, and Prabhat N. Jha, Endophytic Nitrogen-Fixing Bacteria as Biofertilizer, In book: Sustainable Agriculture Reviews 11, Chapter: Endophytic Nitrogen-fixing bacteria as biofertilizer, Publisher: Springer Dordrecht, The Netherlands, Editors: Eric Lichtfouse, pp.183-221)

Quantification of Nitrogen Fixation

Quantification of fixed nitrogen made available to the plants by diazotrophic bacteria can be estimated by the following methods.

i. Total N Difference Method

ii. Acetylene Reduction Assay

iii. Analysis of N Solutes in Xylem and Other Plant Parts

iv. N-Labeling Methods

Total N Difference Method

It is the oldest and simplest method which is based on an assumption that the control plant and the infected plants absorb same amount of N from soil. In this method, there is measure of biological nitrogen fixation on the basis of difference between the total N content of crop grown in presence of diazotrophic bacteria and their counterpart grown without bacterial inoculation.

Acetylene Reduction Assay

Based on the fact that the nitrogenase enzyme involved in N_2 fixation can also reduce acetylene into ethylene as both nitrogen and acetylene are triple bonded structures. This technique involves incubation of bacterial culture with acetylene (0.03–0.1 v/v) in an air-tight vessel. As a result of reduction of acetylene by nitrogenize using a gas chromatograph the gas phase of sample is analyzed for ethylene generation.

The calculated ethylene amount can either be converted into amount of nitrogen fixed by directly multiplying it with a factor 3 or the ethylene amount can either be directly used to quantify the amount of nitrogen fixed. Because to convert N_2 to NH_3, three pairs of electron are used while in acetylene to ethylene conversion only single pair of electrons is used. The acetylene reduction assay is simple, cheap and sensitive technique because this technique measures N_2 - fixation indirectly on the basis of electron flux through nitrogenize.

Analysis of N Solutes in Xylem and Other Plant Parts

The composition of nitrogenous compounds present in plant the xylem sap is analyzed in this method. The basic principle is to differentiate between the nitrogen fixed and the soil derived nitrate N in the plant. In this method, the composition of nitrogenous compounds present in plant the xylem sap is analyzed. The basic idea is to differentiate between the nitrogen fixed and the soil derived nitrate N in the plant. As the uptake from soil increase, it inhibit the N 2 - fixation, which results in the change of composition of nitrogenous compounds in xylem sap and by monitoring such changes, the N 2 - fixation can be quantitatively analyzed.

N-Labeling Methods

There are three different methods for labeling viz.

i. 15 N gas labeling

ii. Isotope dilution

iii. A-value method

The principle behind this method is based on the difference of 15 N/ 14 N present in soil or plant system with that of atmosphere (0.33%). So, if the plants are incubated with 15 N and then, the nitrogen fixation is evaluated in plants, the ratio would differ in plants than that of present in atmosphere. And the changeoccurring in the ratio can be monitored and analyzed for quantifying the nitrogen fixing ability of microbes.

Molecular Analysis of Nitrogen Fixation

Since, the majority of microorganisms cannot be cultured on media, bacterial communities show immense phenotypic and genetic diversity estimation and analysis of natural diazotrophic bacterial communities is quite challenging. However, by employing cultivation independent techniques using universal primers for amplification of gene encoding the key enzyme nitrogenase, this problem can be overcome. There are three types of nitrogenases based on the presence of core metal: (molybdenum(Mo), vanadium (V) and iron (Fe)) which bridge two units of this enzyme. Out of these three types, Mo-nitrogenaseis is the most prevalent.

There are three genes named:

nifHDKwhich encodes for the structural part of nitrogenase complex. Apart from ***nifHDK***, nitrogenase expressionand function depends on several other genes (20 in case of *K. pneumoniae*). The genes

nifHDKencodes for a and b fragments respectively of larger segment of nitrogenase complex called dinitrogenase, while ***nifH***encodes smaller segment Fe protein (dinitrogenase reductase). Nitrogenase activity can be estimated by following procedures.

Plant Growth Promoting Properties of Endophytic Bacteria:

The endophytic bacteria can benefit their host through various growth promoting effects which includes production of phytohormones (auxin and cytokinin), synthesis of siderophore, 1-aminocyclopropane-1-carboxylate-deaminase activity and antagonistic activity, Apart from N2 fixation. To solubilize mineral phosphate, several endophytic bacteria have been reported to have the ability. However, as endophytes reside in the interior of plant issue where insoluble mineral phosphates are not available, this ability may not be useful for plants. The characteristics of some beneficial endophytic bacteria are discussed below.

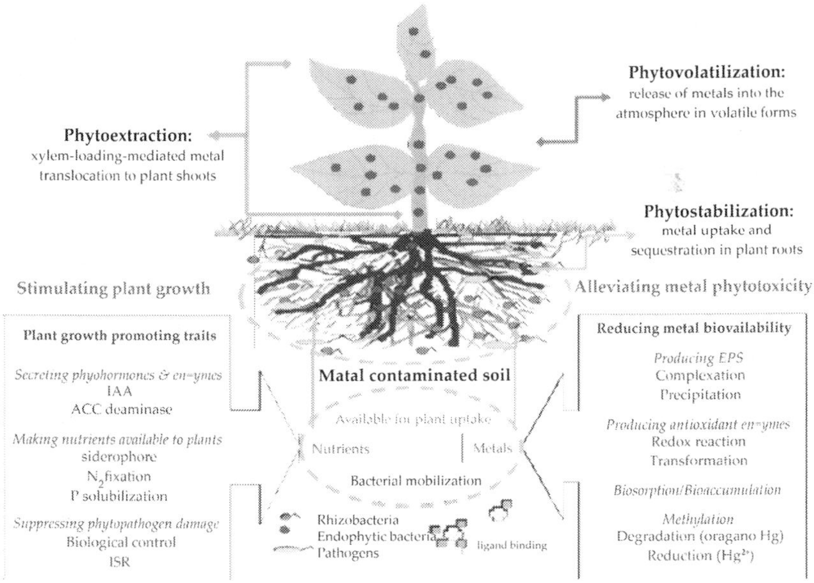

Figure: Plant growth promoting mechanisms of endophytic bacteria in metal contaminated soils (source: Ying Ma, Mani Rajkumar, Chang Zhang, Helena Freitas, Benefical role of bacterial endophytes in heavy metal phytoremediation,Journal of Environmental Management 174:14–25, 2016)

Phytostimulatory Compounds

Plant growth promoting bacteria produce certain volatiles and the co-factor pyrroquinoline quinine. Many plant associated bacteria have been shown to produce auxins chieflyindole-3-acetic acid, which enhances lateral root growth formation and thus, nutrient uptake and root exudation by plants. Most of the beneficial bacteria synthesize indole-3-acetic acid through indole-3-pyruvate pathway. In this pathway, tryptophan to indole-3-pyruvic acid conversion occurs by an aminotransferase, which in turn gets decarboxylated by indole-3-pyruvate decarboxylase to indole-3-acetaldehyde and the oxidation of indole-3-acetaldehyde converts it to indole-3-acetic acid Indole-3-acetic acid synthesis occurs in stationary phase of growth and positively regulated by aromatic amino acid through a regulatory protein *Tyr* R.

As *Bacillus megatarium* UMCV1, Cytokinins are involved in plant growth promotion, a rhizospheric bacterium, can accelerate biomass production in *Arabidopsis thaliana* through the inhibition of primary root growth by allowing increased lateralroot formation and root hair length of host plant.*Methylobacterium* sp. strain NPFM-SB3, isolated from *Sesbania rostrata* is alsofound to produce cytokinin *Azospirillum* sp., *G. diazotrophicus*, *H. seropedicae* to enhance plant growth by producing gibberellin. Even some isolates can produce morethan one phyto-hormone. Isolated*Pantoea agglomerans*, produce four major plant hormonesi.e.abscisic acid, gibberellic acid cytokinin andindole-3-acetic acid. The Identification of the bacterial role for volatile organic compounds can beanalyzied by microarray results and histo-chemical data of *Arabidopsis* seedlingsexposed with *B. subtilis* and stated that the volatile organic compounds can influencethe plant growth by manipulating auxin homeostasis in plants which can be evident frominduction of genes encoding enzymes of metabolism of indole-3-acetic acid.

Induced Systemic Tolerance

In order to face biotic and abiotic stresses like drought, salt, nutrient deficiency or excess, extremes of temperature, presence of toxic metals etc, some of the plant growth promoting bacteria (PGPB) can help the associated plants. For the biotic and abiotic stresses the term used is "induced systemic tolerance" which is generated by PGPB-induced physical and chemical changes in plants. From the production of bacterial 1-aminocyclopropane-1-carboxylate de-aminase activity, antioxidants, cytokinin or volatile organic compounds, Induced systemic tolerance happens. Ethylene is produced by plant to regulate plant homeostasis, by the action of many biotic or abiotic stresses. The production of ethylene is prohibited beyond a threshold level, because it reduces root and shoot development and thus it is described as "stress ethylene". Some endophytes have the property to synthesize 1-aminocyclopropane-1-carboxylate-deaminase, that can degrade the immediate precursor of ethylene from root exudates while converting it to a -ketobutyrate and ammonia and thus, can promote growth of plant in the vicinity.

There are several mechanisms for inducing systemic tolerance generation in response to stresses. The volatile organic compounds mediate salt tolerance thus affecting abscisic acid signaling of plants during stressby the production of cytokinin and through production of antioxidant catalase which are all included in such mechanisms of induced systemic tolerance. The promotion of plant growth has been well described for the role of phyto-hormone produced by associative bacteria during salinity or drought stress.

In drought condition, the production of Indole-3-acetic acid in bacteria stimulates the formation of well-developed roots to provide sufficient water from soil.

Biocontrol Agent

To control diseases, the application of microorganism is one of the most promising way because microorganisms are eco-friendly and cost-effective. The microorganisms have to be stable under varying condition of pH, temperature and concentrations of different ions, in order tobecome an efficient bio-control agent. After reducing the growth of pathogenic organisms at laboratory, greenhouse/ field level in various studies, many endophytic bacteria are supposed to benefit host plant. An efficient biocontrol agent must also possess certain traits like:

- Efficient colonizer of root to deliver antibiotic along the whole root system.
- To protect itself from predators (protozoans) in the rhizosphere.
- Release antibiotic in right microniche.

Bacteria limit pathogen indirectly by inducing systemic resistance in plants against certain diseases or the systemic resistance is directly induced through antagonistic property, competition for iron, detoxification or degradation of virulence factors. Endophytic bacterial bio-control agents inhibit the growth of fungal/ bacterial pathogens by different mechanisms, some of which are described below.

Antagonism

By the production of allele chemicals/antibiotics, Endophytic bacteria exhibit bio-control activity including both antifungal and antibacterial. Gram negative bio-control agents like Pseudomonas produce HCN, pyoleutorin, pyrrolnitrin, 2,4-diacetylphloroglucinoland phenazines chiefly phenazine-1-carboxylic acid and phenazine-1-carboxamide. In Pseudomonas fluorescens CHA, all the above mentioned metabolites are helpful for biocontrol. But, the expression of their genes can change with different plant cultivars hexyl- 5-propyl resorcinol produced by the antagonistic bacterial strains has also been demonstrated in last couple of years. Munumbicin, an antibiotic produced by endophytic bacteria inhibits growth of phytopathogenic fungi P.ultimum and F. oxysporum. Certain volatile organic compounds like 2,3-butanediol, or blends of volatiles produced by Bacillus sp. also act as biocontrol agents. Nutritional factors i.e. type of carbon source utilized, trace elements and availability of other nutrients as well as non-nutritional factors like environmental influencies level of antibiotic synthesis

5
CHAPTER

NON-SYMBIOTIC NITROGEN FIXERS

AZOSPIRILLUM

About Eighty percent (80%) of the atmosphere is nitrogen gas N_2 but it cannot be utilized by most living organisms. Due to less quantity of nitrogen, Plants, animals, and microorganisms die when they are surrounded by N_2 they cannot use. Among the Nitrogen-containing components necessary for life is ammonia (NH_3) which is another form of nitrogen to manufacture amino acids, proteins, nucleic acids. Biological nitrogen fixation (BNF) process changes inert N_2 to beneficial NH_3. Thisnitrogen fixation process is facilitated by bacteria and certain species of actinomycetes in nature. Plants get benefit when the bacteria die and release nitrogen to the environment, or when the bacteria are loosely associated with the roots of plants. The bacteria live in small club-like growths on the roots called nodules, in legumes and in a few other plants. N_2 fixation occurs within these nodules, and the NH_3 produced is directly absorbed by the plant. *Rhizobium* bacterium and a legume host plant have close/symbiotic relationship in between due to this relationship Nitrogen fixation occurs in legumes.

In nature, the Biological nitrogen fixation takes many forms:

- Blue-green algae which includes blue green algea (nostoc)
- Non-legume ,Lichens and actinomycetes,
- Associative symbiosis
- Free-living bacteria in soil

A significant quantity of $NH_3$is contributed to natural ecosystems by nitrogen fixation. In a natural ecosystem, between the ranges of 25-75pounds of nitrogen per acre per year is fixed legumes by Nitrogen fixation phenomena and several hundred pounds of nitrogen in a cropping system. Through associations between plant roots with nitrogen-fixing bacteria, symbiotic N_2 fixation occurs.

The symbiosis is between two species, like:

- The formation of nodules on roots due to the symbiotic relation between *Rhizobium* or *Bradyrhizobium*and leguminous species.
- The symbiotic relationship between *Frankia* anda small numbers of non-leguminous genera is also reported beneficial.

The BNF technology between cereals crops and grass is beneficial for the raising of economical agriculture benefit. Several findings deduced results that associations between NFB (nitrogen fixing bacteria) and tropical grass results in contributing a beneficial N fixation to these plants under favorable conditions.

Where fertilizer inputs are undesirable, not possible or impractical, the use of *Azospirillum* as a "biofertilizer" is important in agricultural systems. For enhanced plant growth, the inoculation of *Azospirillum* or other diazotrophs on crops can result in enhanced nitrogen content under environmental conditions. As this inoculation changes nutrients adjustments and modify the functioning and size of roots. In inoculated plants, there was observed high N_2 fixation as compared to the results of a study in which inoculation was not practiced. In a study, it was found that the nitrogen fixation rate is high near the flowering stage under good soil moisture and high temperature conditions.

Benefits of Azospirillum Inoculation:

The inoculation of *Azospirillum other than to enhance nitrogen, has many benefits which are as* following:

- development, branching and Promotion of root hair
- the uptake of microelements and N, P, K upto Increased levels
- good water status in plants
- In grain crops the presence of sufficient dry mass buildup, hence better yield.

The invasion of the cortical layer was found by the Inoculated plants when they were examined under the electron microscopes. The isolated nitrogen fixing bacteria from native grass (*Saccharum spontaneum* L.) root were used as inoculant and their capability was tested to see increase shoot growth and root density in corn, sugarcane and rice yields and the results as expected were good. The bacteria were found to possess at least 57 % of the characteristics of the genus Azospirillum. In different regions, the field tests result reports assured the beneficial role of nitrogen-fixing bacteria in the enhancement of crop yield of few vegetable, rice and corn thus improving their overall yield. For good stimulated plant growth, the associative N_2 fixing bacteria were found capable of gibberellins and cytokinins production which are growth regulators in plants and stimulate plant growth.

Those bacteria were from *Azospirillum genus which is the most* promising microorganism for many cereal crops and grass.

Azospirillum species are described as Gram negative, rod-shaped, 1mm in diameter and very motile. Its cells are about 1.0 um x 3.5 mm in size single flagellum when grown in MPSS broth while lateral flagella when grown on MPSS agar at 30 °C. They also form wrinkled, dark pink colonies when grown on MPSS agar. A formation of a white veil or bacteria band, is visible when inoculated into an Nfb and Dobereiner's liquid medium. *Azospirillum* utilizes glucose, lactate, succinate, fructose, malate, pyruvate, fumarate, as carbon source, reduced nitrate and does not require biotin. *Azospirillum* species are motile, rod-shaped of 1mm in diameter and described as Gram negative bacteria. Their cell size is about 1.0 um x 3.5 mm. When flagellum is grown in MPSS broth and also at 30 °Clateral flagella grows on MPSS agar. Dark pink, wrinkled colonies were formed on MPSS agar. A bacteria band or white veil formation can be so visible when it's inoculated on Dobereiner's liquid medium and Nfb. As carbon source, glucose, lactate, succinate, fructose, malate, pyruvate, fumigate are used by *Azospirillum. It* does not require biotin and utilize, reduced nitrate.

The N sources used by *Azospirillum* for their growth are:

i. Ammonium

ii. Nitrate

iii. Amino acids

iv. Elemental N

Azospirillum spp. being able to grow under following conditions and are highly adaptable:

i. Anaerobic conditions (nitrate used as eletron acceptor)

ii. Microaerobic (elemental or ammonia used as N source)

iii. Fully aerobic conditions (ammonia, nitrate, amino acid or combined N only)

Associative Nitrogen Fixing Bacteria:

Associative nitrogen fixing bacterium contributes passively to the plant growth by fixing N_2 from the atmosphere and lives on rhizospherial environments. in rice rhizosphere, The common associative N_2fixing bacteriaare *Alcaligenes faecalis, Enterobacter cloacae, Klebsiella oxytoca, Klebsiella planticola, Azospirillum brasilense*and *Azospirillum lipferum*.Some strains have been isolated from *Alcaligenes faecalis*and *Azospirillum brasilense.*But only safe strains and proved bio-fertilizers are *Azospirillum,Enterobactercloacae, Alcaligenes faecalis*and *Klebseilla pneumonia. In* paddy fields, the number of associative N_2 fixing bacteria is more than that in dry fields as large as 103-107 cells/g soil.

The Associative N_2 fixing bacteria's place of inhabitance is rhizosphere.

Autotroph

- The Photosynthetic N-fixing bacteria ***Rhodobacter(rhodopseudomonas) Rhodospirillum***

Heterotroph

- Autofixing or aerobic Bacteria are named ***Azotobacter, Azotomonas***
- Autofixing or slightly aerobic Bacteria are named ***Derxia, Methylomonas***
- Autofixing or anaerobic-aerobic Bacteria are named ***Bacillus***
- Autofixing or anaerobicBacteria are named*Clostridium,* ***Desulfotomaculum, Desulfovibrio***
- Associative nitrogen fixing or aerobicbacteria are named***Beijerinckia***
- Associative nitrogen fixing or slightly anaerobicbacteria are named ***Alcaligenes, Arthrobater, Azospirillum, Flavobacterium, Pseudomonas***
- Associative nitrogen fixing oroxidative-reductivebacteria are named ***Enterobacter, Klebsiella.***

The chemical characteristics of root residues and components of root exudates define the types and number of soil microorganisms present. The carbon sources that are required for the survival of rhizobacteria are supplied by plant roots because the rhizobacteria has such a weak capacity of degrading organic matter. In addition to that, there are also some growth regulators present in root exudates antibiotics which regulate the growth of associative N_2fixing bacteria. To study the effect of plant roots those have selectivity for microbes' types many studies were conducted. One strain of associative nitrogen fixing bacteria was inoculated to different varieties of rice and difference in nitrogen fixation activity was found. The associative nitrogen fixing bacteria along with sugars for carbon source and organic acids and GA3 growth regulators were supplied by Rice root exudates and residues. The Associative N_2 fixing bacteria belong to PGPR (Plant Growth-Promoting Rhizobacteria) either the test results show any beneficial or adverse effect of rhizobacteria.

- The Nitrogen source of ammonia exudates can provide to rice by Associative N_2 fixing bacteria.
- The exudation of growth regulators such as GA3 and IAA enhance the growth of rice roots and also provide Associative N_2 fixing bacteria.
- Most associative N_2 fixing bacteria have nitratase, which goes into plant roots after inoculation and assist in nitrate reduction in plant and increase the N level, hence enhancing N_2 fixation.
- The plant mineral uptake can also be enhanced by Associative N_2 fixing bacteria.
- For the enhancement the growth of lateral roots, Associative N_2 fixing bacteria can be used.

ISOLATION OF MICROBIAL STRAINS

Isolation of endophytic bacteria from grass/rice roots

There are several methods designed to isolate the nitrogen-fixing bacteria from rhizosphereand from grass hosts roots. There are following techniques that are used to isolate nitrogen-fixing bacteria. Field grown roots of intended hosts or its relatives are the target materials. Any of the fallowing methods can be used:

- From the diverse soil environment, the collection of roots from healthy looking host plants. The roots are to be placed in clean plastic bags and to be sealed. It is good to keep bag inside Styrofoam container provided with small amount of ice to keep temperature from rising while in transit.

  i. The young and mature roots selection.
  ii. The use of 50 % laundry bleach for surface sterilizing
  iii. The use of sterile distilled water, 3-5 times washing.
  iv. Dividing the roots into young and mature and further dividing each of them into 3 sets;

a. One set to be kept intact and

b. The other 2 for grinding.

Set (A) Use of whole/Intact roots:

To isolate for root endophyte, it is necessary to surface sterilize the roots. When isolating for general population of root associated bacteria, surface sterilization of the roots is not necessary. Sterilization can be roughly conducted by slowly shaking the roots for 20 min. in 10-15 % chlorox bleach (by volume) followed by 3 times rinse in sterile water.

- Cutting roots into 1-2 cm length.
- Inoculating into N-free defined semi solid medium.
- Incubating at 28-30 °C for 2-3 days; observe for pellicle formation below surface of medium.
- Using a wire loop, getting sample from tubes with good growth.
- Streaking onto N-free defined agar medium.
- Incubating plates inside at 28-30 °C.
- Examining for colonies every day, picking individual colonies when fully grown.
- Transfering individual colonies into semi-solid tubes.
- Running ARA on individual tubes.
- Marking tubes with high nitrogenase activity and setting aside for purification and identification.

Set B Crush roots

Dividing into 2 portions; one portion to be used in stab culture (B-1), the other as inoculum for rice or intended trap host (B-2).

Set B-1 Use of crushed roots

- By means of a long fine forcep or wire loop getting sample from grind roots and stabbing into semi-solid N-free defined medium.
- As in A, observing for growth of pellicle below surface of medium daily

- With the use of a wire loop obtaining sample of the pellicle and streak into plates of N-free medium.
- Running ARA as in A and compare nitrogenize activity
- Keeping tubes with high nitrogenize activity and set aside for purification, identification and further evaluation.

Set B-2 Use of trap host

i. Preparation of N-free liquid mineral medium suitable for the test plant e.g. rice.

ii. Sterilization of dehulled seeds.

iii. Pre-germination in sterile agar plates.

iv. When radicle/embryonic roots come out, selection of ones that do not show contamination.

v. Carefully transfering loopfuls of the colonies obtained from grind preparation from B into each of the tubes containing semi-solid medium.

vi. Transfering a germinant into the semi inoculated tube.

vii. Collection of the roots from tubes in # 8 and add a drop of sterile water.

viii. Streaking loopfuls of juice from #9 into N-free agar plates containing defined medium.

ix. Picking individual colonies and transfer individual colonies into tubes containing N-free semi-solid medium.

x. Incubation and when obvious growth develops, running ARA again.

xi. Separation of tubes with high ARA and setting aside for future purification and identification.

Cautions and Limitations of Biofertilizer

i. Never to be mixed with chemical nitrogen fertilizers;

ii. Never to apply with fungicides, plant ash etc. at the same time;

iii. Never directexposure to sunlight;

iv. Not keeping used solution overnight;

v. Storing at room temperature, not below 0°C and over 35uy°C.

Biofertilizer with associative N_2 fixing bacteria act as supplement for nitrogen requirement of corn, rice and sugarcane. It is essential to apply 30-50 % of the recommended inorganic forms of nitrogen to meet the requirements for other nutrients such as phosphorous and potassium.

For sustainable agriculture, microorganisms are suitable and are the potential tool for future. In order to reduce external nitrogen input and improvement in the quality and quantity of internal resources, the BNF process are economically attractive and ecologically good. For maximum bacterial strains, research should be conducted to know the bacteria that are useful and necessary for different environmental conditions. In this regard, field experiments are needed to understand better biological efficacy for increased yields in crop systems. The understanding of symbiosis in the plant family can be enhanced by the availability of complete genome sequences and also by functional genomics of symbiotic microorganism's bacteria and mycorrhizal fungi and to get this knowledge is a big challenge for future research.

To meet the current challenge of fast-growing worldwide demand for agricultural products, the increased use of BNF is one of the major pathways to maintain/increase yield and to reduce the environmental footprint of agriculture.

Biological Nitrogen Fixation: Importance, Associated Diversity, and Estimates

In nutrient cycling, many soil microorganisms play an important role. Among the biological nitrogen fixation (BNF) ***diazotrophs bacteria*** are one of them. These bacteria are associative with plant species, free-living or are established symbiose with legumes. Due to their contribution in nitrogen supply, the nitrogen supply in different ecosystems, including natural and managed systems reveal the great importance of diazotrophic organisms. It is estimated that global BNF contributes about 122 Tg of Nitrogen yearly with cultivated agricultural systems fixing from 33 to 43 Tg, that occurs mostly by legume-rhizobia symbiosis. There is a large potential of BNF contribution by associative systems with tropical grasses, but there is uncertainty in these estimates due to several assumptions in the estimation process and fewer studies with this system when compared to the legume-rhizobia symbiosis. As reliable components in the management of sustainable agricultural systems, application of these biological systems can be facilitated for their development in terms of understanding colonization ability, action mechanisms, diversity, formulation, and application of these biological systems can be a recent progress. To develop commercial inoculants using these organisms there areseveral efforts made. The latest progress in using microorganisms that fix nitrogen in a variety of applications is summarized and discussed herein.

In nucleic acids, nitrogen is a very important component, proteins and other nitrogen compounds and because of that microorganisms that carry out biological nitrogen fixation (BNF) have much importance and so the nitrogen is an essential component of life for all living being. The symbiotic nitrogen-fixing bacteria with legume species are commonly known as alpha and beta rhizobia that perform BNF mechanism and provide high sustainability for ecosystems. The microorganisms help promote plants growth by production mechanism of siderophores, exopolysaccharides (EPS), and phytohormones; phosphate solubilization and protection against phytopathogenic fungus.

In diverse habitats, Diazotrophs are commonly found in symbiotic associations as in termite guts, free-living in soil and water, associative symbioses as in grasses, actinorhizal association especially found in woody plants, cyanobacterial symbioses plants as well asroot-nodule symbioses with legumes. Nitrogen fixation, acquisition of Phosphorus and other nutrients by mycorrhizae are major types of symbioses. For legumescultivation, relationship of rhizobia & mycorrhiza are of great importance as they can influence the infection rate, mineral nutrition and also physical and chemical conditions of the soil as well by the addition of organic waste that increases the growth of plants Symbiotic & mutualistic relationships involve rhizobia which depends on chemical signals between two organisms. Selection of maximum combination of the rhizobia and host results in more effective symbiosis & better growth of the host plant.

In addition to the other microorganisms there are several Biofertilizers that have ability to cater different needs of growing plants act as a consortium in the rhizosphere, there is clarity found about the interaction between the consortium of microbial inoculants & plant systems providing the foundation for coupling more advantages from microbial inoculants to improve plant growth yield. There are many detailed literature articles to discuss the operational mechanisms of PGPB/PGPR that have been published in past few years. It is utmost important to produce microbial inoculants formulation step and to determine the success of a biological agent. In last few decades, the biopolymers act as potential inoculants and the studies have demonstrated them as best potential for bacterial carriers for microbial inoculants.

Key for Ecological Success-BNF

As everyone knows that Nitrogen is one of the most abundant and is very essential macronutrient for plant species. Many bacteria can reduce $N_2$by the help of enzymes and turn it into ammonia. And this ammonia is used in the synthesis of essential elements and this process called Biological Nitrogen fixation. It is symbiotic when there is a mutualistic association between fixing microorganisms which are mainly rhizobia and the plant species. It can be symbiotic also when carried out by free-living fixing microorganisms.

According to the morphological, genetic, physiological and the phylogenetic characteristics, there is a wide distribution of rhizobia in different taxonomic groups. Rhizobia are classified in two types according to most recent studies;

- α- -rhizobia in which the genera *Agrobacterium* , *Allorhizobium* , *Azorhizobium* , *Bradyrhizobium* , *Devosia* , *Mesorhizobium* , *Methylobacterium* , *Ochrobactrum* , *Phyllobacterium* , *Rhizobium* , and *Sinorhizobium* are included and
- β-rhizobia in which bacteria of the genera *Burkholderia* , *Cupriavidus* , and *Herbaspirillum* are associated

There are benefits of legume-rhizobia in symbiotic system and also in the development of nodules which results inexchange of molecular signals of chemical nature between plant and its symbiont. It is a known thing that legumes form root nodules in response to Nod factors and the plant supply endogenous signals, particularly called plant hormones, which is considered important for the formation of good symbiotic interactions between legumes and rhizobia. The nitrogen-fixing bacteria fromnative species perform nitrogen fixation at low degrees. Because of this behavior, it is important to obtain efficient BNF for the collection of elite strains of rhizobia. In order to reduce one mole of nitrogen, there re 16 mol of adenosine triphosphate (ATP) required by Microorganisms to fix nitrogen.

The BNG bacteria get energy through oxidation of organic molecules. The free living, Non-photosynthetic microorganisms can obtain molecules from different photosynthetic microorganisms, just like cyanobacteria utilize sugar molecules produced by photosynthesis. Another example is symbiotic &Associative nitrogen-fixing microorganisms which obtain energy compounds from the rhizosphere of their host plants.

The increase in BNF usage can be a major source of nitrogen for plants which is beneficial for the Progress in agricultural sustainability. The agricultural systems should use effective &long-term sustainable management of internal resources. Thus it can be said that BNF is economically attractive and ecologically good for reducing external nitrogen input and to improve the quality & quantity of resources.

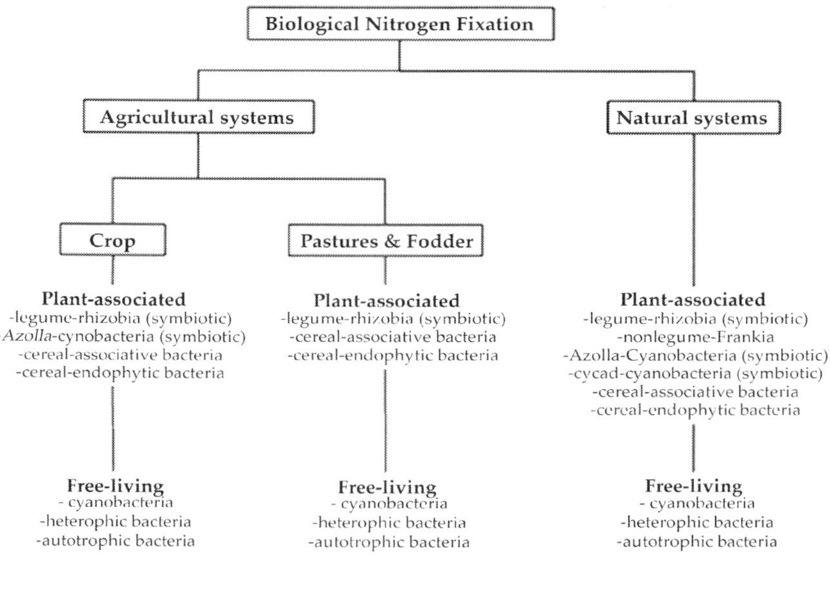

Figure: Biological Nitrogen Fixation

Clearly it is false to consider sustainable agriculture without BNF on a big scale. To maximize the contribution of BNF for sustainable agriculture, more research is required. The efficient use of Nitrogen yields many advantages e.g. genetic improvements to adapt plants to nitrogen-poor soils and the reduction of fertilizers in order to maintain productive levels high. Nitrogen is used& obtained through assimilation, redistribution of itself and absorption in cells as well as keeping the good balance between nitrogen storage & use. To increase the use &efficiency of nitrogen, there are new methods designed that are being intensely studied. The new methods include molecular & biochemical pathways of assimilation and absorption in plants. Agro-ecological methods like BNF allow sustainable use of nutrients without production loss.

Bacterial Biofertilizers:

Plant yield is always dependent on nutrients availability, such as nitrogen, and farmers apply 100 kg of N per hectare at least to achieve the goal. But Nitrogen fertilizers are less expensive & chemical fertilizers produce soil pollution. The bio-fertilizers are now becoming popular for sustainable agriculture. "biofertilizer" specifies that the fertilizer contains all nutritional requirements that are essential for a crop through microbiological means. In many countries, biofertilizers are also called "bacterial inoculants".

A bacterial inoculant is a good assistance for bacterial strain or species in an easy-to-use &economical carrier material. Inoculants are the "carriers" which ride on themselves the living bacteria and take them to the living plants to produce the desired plant growth. With rhizobialegume seeds are inoculated, BNF are used for the growth of legume plants. In the past years of 1950, the utilization of inoculants for legume plants started in Brazil and now in this era it has been stated again, the inoculation process saves about 11 billion dollars per year for a single soybean crop only. Although, the research and production of inoculants for legumes has a long tradition in Brazil. In recent studies, the nitrogen fixation in *Azospirillum* grass has just used; hence, there are few commercial inoculants that are available in the market for commercial purposes. Peat has high water-holding capacity & large surface area, due to these properties it is the most frequently used carrier for the rhizobial inoculant industry. It supportsrhizobial growth and survival in huge quantity. However, a noteworthy processing e.g. drying, mining, milling, and neutralization are beneficial for peat-based inoculant before using it in a commercial production system. The formulation steps are crucial aspect for producing microbial inoculants and they determine the success of a biological agent. The peat-based inoculant formation createsviable bacteria on a suitable carrier. The additives are also added in it which aid in the protection &stabilizationof microbiological cells during the transportation & storage until they have reached safely to the target place. The formulationsare easy to apply and handle so that their delivery to the targets is safe and formulations should also protect bacteria from harmful environmental factors by maintaining/ enhancing the activity of the organisms in the field

Features of a superior carrier

A superior carrier material should be present in a good quality inoculant. The characteristics of a superior quality carrier material for microbial inoculants must have following features shown in figure below:

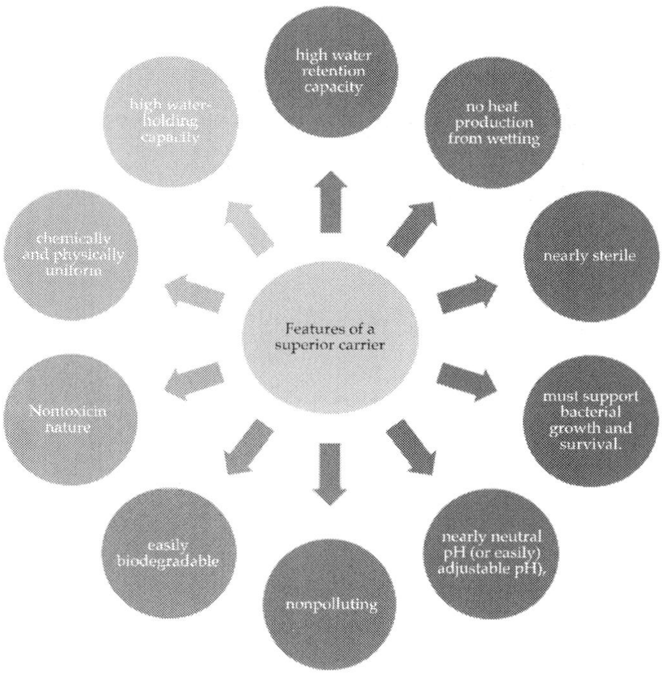

Figure: Different features of a superior carrier

Physical Availability of Inoculants

Inoculants can be purchased in four basic dispersal forms i.e. slurries, powders, liquids and granules. It is always a farmers' choice to use each type of inoculant depends upon the need, cost and market availability also on the need of a particular crop under certain environmental conditions.

There are new carrier materials researched that have also been available naturally, these materials include inorganic soils compost, wheat bran and clays, spent mushroom compost and spent agricultural waste materials. There are many other inert and synthetic materials e.g. perlite, ground rock phosphate, vermiculite, polyacrylamide gels, calcium sulfate, &alginate, areanalyzed. In the processing of solid carriers there are many problems associated with them. The Liquid inoculant formulations production is a solution. Liquid inoculant formulations use different broth cultures amended by materials which promote cell survival in the packaging and maintenance of cell survival after application to seeds and soil. The Additives to liquid inoculant formulations protects the cells on seeds during desiccation high temperatures.

For many inoculants production, many types of biopolymers can be used that have good rheological properties, because of their ability to limit heat transfer &high water activities. In last few years, biopolymers as inoculants are studied as being strong potential. Biopolymers are polymers of proteins, nucleic acids, or polysaccharides produced by living organisms. These polymers are defined as potential bacterial carriers for microbial inoculants. Living cells are encapsulated in these formulations thus protecting microorganisms from many environmental stresses, and gradually release the cells into soil in huge quantities.

Biofilm is also another recent possibility for the development of new inoculants. The use of Biofilm consists of microbial cells of algae, bacterial & fungi along with other microbial cells. In addition to an extracellular biopolymer also known as extracellular polymeric substance produced by the cells, as compared to monocultures (EPS) provides structure &protection to the plants community. The formation of fungal-bacterial bio-films (FBBs) increases metabolic activities through bacterial colonization on biotic fungal surfaces. The addition of nitrogen N_2fixing rhizobial strain to FBBs to make fungal-rhizobial biofilms (FRBs) thus improves potential biofilm applications in nitrogen deficient situations and in the production of biofilm inoculants for bio-control in plants. Biofilms help in the nutrients cycling attached to the plant roots of few crops and also in the biocontrol of diseases &pests. This nutrientscycling show upgraded agricultural productivity results in the end. A developed FRB inoculant gives significantN_2 fixation in soybean by approximately 30 % as compared to some conventional inoculant thus consisting of monoculture inoculant i.e. only rhizobia. The reports indicate that these symbiotic bacteria can be used as PGPR with non-legume plants. The heavy colonization of FBBs/FRBs on root hairs of tea "*Camellia sinensis*" rice "*Oryza sativa*", wheat "*Triticum aestivum*"& anthurium "*Anthurium andraeanum*", and FRBs act as "pseudonodules" by fixing N_2on the roots of nonlegumes biologically.

6
CHAPTER

PHOSPHORUS BIOFERTILIZERS

INTRODUCTION

Phosphorus solubilizing bacteria (PSB) play an important role in phosphorus nutrition by enhancing its accessibility to plants through solubilization and mineralization of inorganic and organic phosphates in soil. The insufficient phosphate availability limits the growth and yield of plants and therefore, lesser solubility of hydroxyapatite, calcium triphosphate and aluminum phosphate resulted into low accessibility of phosphate in agricultural soil as well as is the plants. Phosphorus is an important limiting factor in agriculture production and considering the negative effects of chemical P fertilizers, microbial intervention of PSM seems to be an effective way to solve the phosphorus availability in soil. PGPR with phosphate solubilization, pesticide degradation potential and ability to excrete phytohormones and antimicrobial metabolites may be used as potent biological inoculants for sustainable agriculture. Now it is evident that biological methods demonstrate a number of economic and environmental advantages over chemical-based P fertilizers applications. Numerous evidences indicate that PSB pose wide variety of functions which facilitate their implications in large number of agro-, bio-, and environmental technological strategies.

Phosphate solubilizing bacteria (PSB)

Phosphate solubilizing bacteria are helpful bacteria for solubilization of inorganic phosphorus from insoluble compounds. One of the most important traits associated with plant phosphate nutrition is P-solubilization ability of rhizosphere microorganisms. It is generally accepted that the mechanism of mineral phosphate solubilization by PSB strains is associated with the release of low molecular weight organic acids, through which their hydroxyl and carboxyl groups chelate the cations bound to phosphate, thereby converting it into soluble forms. As phosphate Biofertilizer, PSB have been introduced to the Agricultural community. Phosphorus "P" is a major macronutrient and essential for plant growth. In the form of phosphate fertilizers, It is applied to the soil. However, a large quantity of soluble inorganic phosphate can be applied to the soil as chemical fertilizer which becomes immobilized quickly and becomes unavailable to plants. In a present scenerio, the priority is to manage soil phosphorus to optimize crop production and minimize P loss from soils.in agriculturists, PSB is an attractive soil inoculums to improve the plant yield and growth., about 50% of the crop requirement can be saved When PSB is used with rock phosphate of phosphatic fertilizer. The "P" uptake by plants is increased by the use of PSB as inoculants. Simple PSB inoculation of seeds gives crop yield equivalent to 30 kg P_2O_5 /ha which is 50 percent of the need for phosphatic fertilizers.

- Through fertigation or hydroponic operationsPSB can be applied, Alternatively. Many different strains of PSB are *Microbacterium laevaniformans* (P7) *Pseudomonas putida* (P13) and *Pantoea agglomerans* (P5), these strains are highly efficient as they are insoluble phosphate solubilizers. Researchersat Colorado State most recently demonstrated consortia of four bacteria which sold commercially as Mammoth P University which are synergistically phosphorus soluble at a much faster rate than any single strain alone.
- The modern agriculture is getting more and more dependent upon the steady supply of synthetic inputs with the introduction of green revolution technologies. Manifold increase in the

productivity of farm commodities is due to Intensive agriculture with the use of large amount of chemical fertilizers. The harmful effect of these chemicals are clearly visible on, micro flora, soil structure, , food and quality of water and fodder. At this critical juncture, biofertilizers are useful supplement to chemical fertilizers.

With all the nutrients and live microorganisms required for the benefits of the plants, biofertilizers are natural and organic fertilizer that helps to keep in the soil. in organic farming It is used as sustainable agriculture for farming green and non-pollution crops. As they are renewable source of plant nutrients to supplement the chemical fertilizers, cost effective and for sustainable agriculture they are also one of the important components of integrated nutrient management.

The aim of bio-fertilizers is to complement and replace conventional chemical fertilizers where appropriate, so that their use can be reduced with environmental and economic benefits. Biofertilizer are when applied to soil contains living microorganisms; they colonize the rhizosphere or interior of the plant that promotes growth by increasing the availability of primary nutrients to the host plant.

They encourage plants to absorb, a greater quantity of nutrients on their own, which even if they are naturally present in the soil cannot be assimilated by plants due their presence in insoluble form. However, Conventional chemical fertilizers supply the soil with chemical elements despite functioning as a fertilizer. They end up contaminating aquifers if they are not managed at the right moment and in the right dose. Fungus penetrates the cortical cells of the roots, increase surface area of roots, displace of absorption equilibrium of phosphate ions which increases the transfer of P ions, stimulate metabolic processes, Arbuscles absorb these nutrients into the root system

Environment friendly Phosphorus Biofertilizer- Alternative to Chemical Fertilizers

Plant growth promoting rhizobacteria (PGPR) are a group of soil microorganisms which can enhance plant growth parameters and can be used as biofertilizers. The aim of this chapter is to emphasize on the importance and use of microbial inoculants possessing P-solubilizing activities as an environment-friendly alternative to chemical phosphorus (P) fertilizers in agricultural soils. The reckless and unbalanced use of chemical fertilizers and pesticides in agricultural soil has resulted not only in the deterioration of soil health but also created some major environmental and health related issues such as soil and water pollution. Phosphorus is an essential macronutrient required for every aspect of cell biology from energy metabolism to the structure of the genetic material, and its deficiency is a severe constraint to crop production. Soil microorganisms that solubilize the least mobile P play very important role in soil biology, plant health and subsequent enhancement of crop yield. To avoid the phosphorus deficiency, phosphate solubilizing bacteria (PSB) play an important role in supplying phosphate to plants in a more environment-friendly and sustainable manner. Therefore, PSB can be used in an ideal cropping system with a lesser impact on the environment through decreased application of chemical fertilizers.

Soil sustains an immense diversity of prokaryotic and eukaryotic organisms; it is the natural growth media for living plants and microbes. Rhizosphere is niche for a group of soil microorganisms inhabiting around the plant root zone. Basically, rhizosphere is itself. The term "rhizosphere" was coined by German agronomist Lorenz Hiltner in 1904 and is defined as biologically active zone of soil in the immediate vicinity of plant roots, an extremely important area for root metabolism, where microbes flourish and accelerate by root exudates and rhizodeposits. The soil particles strongly adhere to plant rootsurfacescalled rhizoplane, and the root itself is a part of the system, because certainendophytic microorganisms are able to colonize inner root tissues; the stimulation of microbial growth by root exudates is called rhizosphere effect; and it is a dynamic process initiated by root exudates and release of organic nutrients and is influenced by host factors (stage of development, species and age), and soil factors (moisture level, fertility, physical properties and environmental factors such as light and temperature).Generally, rhizosphere is a dynamic environment, which harbours diverse group of microbes in which the effects of the plant on soil microorganisms and effects ofmicroorganisms on plants are interacting and interdependent. The plant-associated bacteria that are able to colonize and persist on roots are called rhizobacteria. The Rhizosphere harbours numerous microorganisms such as algae, bacteria, protozoa and fungi which coexist in the rhizospheric region; bacterial population is most abundant among them. These microbial communities in nature tend to live in relative harmony where all microbial populations generally balance each other out in their quest for food and space.

Plant Growth Promoting Rhizobacteria (PGPR)

The term PGPR was coined by Kloepper and Schroth in late 1970s to define microbial population in the rhizosphere which is naturally occurring, actively colonize plant roots, and shows plant growth promotion activity. They are the potential tools for sustainable agriculture and trend for the future. The plant roots basically provide niche for the proliferation of soil bacteria that thrive on root lysates and root exudates. The population densities of bacteria in rhizosphere is 100-fold higher than in bulk soil because of the high nutrient levels in rhizoshere; and up to 15% of plant roots are covered by micro-colonies of a variety of bacterial strains Bacterial species belonging to genera *Pseudomonas, Bacillus, Rhizobium, Azospirillum, Arthrobacter, Burkholderia, Serratia, Flavobacterium, Enterobacter, Erwinia, Acinetobacter, Alcaligenes,* etc. are well known for their growth promotion effects on plants.

PGPR can be divided into two groups on the basis of their relationship with the hostplants:

- Symbiotic rhizobacteria, which may invade the interior of cells and survive within the cell (also called intracellular PGPR (iPGPR), e.g., nodule bacteria (*Rhizobium*);and
- Free-living or non-symbiotic rhizobacteria that exist outside plant cells (calledextracellular PGPR (ePGPR), e.g., *Bacillus, Pseudomonas, Burkholderia,* and*Azotobacter*).

Characteristics of Ideal PGPR

An ideal PGPR exhibits following characteristics

i. Enhancement of plant growth

ii. Safe to environment

iii. Compatible with other rhizobacteria

iv. High rhizosphere competence

v. High competitive saprophytic ability

vi. Ease for mass multiplication

vii. Broad spectrum of action against pathogens

viii. Tolerant to drought, heat, oxidizing agent and UV radiations.

ix. Excellent and reliable pest control.

Application of Pgpr in Agriculture

Chemical fertilizers are offered with alternatives which are Environment-friendly biotechnological approaches. The excessive use and increasing costs of chemical fertilizers creates anegative impact on agricultural practices, therefore, PGPB is being considered as an alternative way of reducing the use of chemicals in agriculture. During the pastcouple of decades, the use of plant growth promoting rhizobacteria (PGPR) for sustainable agriculture has increased tremendously in various parts of the world. Treatment with PGPR has increased the germination percentage, seedling vigour, emergence, and shoot growth, and root growth, total biomass of the plants, seed weight, early flowering, increased grain, fruit yields and fodder. Significant increases in growth and yield of agronomically important crops in response to inoculation with PGPR have been repeatedly reported. The use of PGPR with the aim of improving nutrient availability for plants is an important practice in agriculture production. According to their mode of action, PGPR have been divided into two broad groups:

i. PGPR that directly affect plant growth, seed emergence or improve crop yields and

ii. Biocontrol PGPR that indirectly benefit the plant growth.

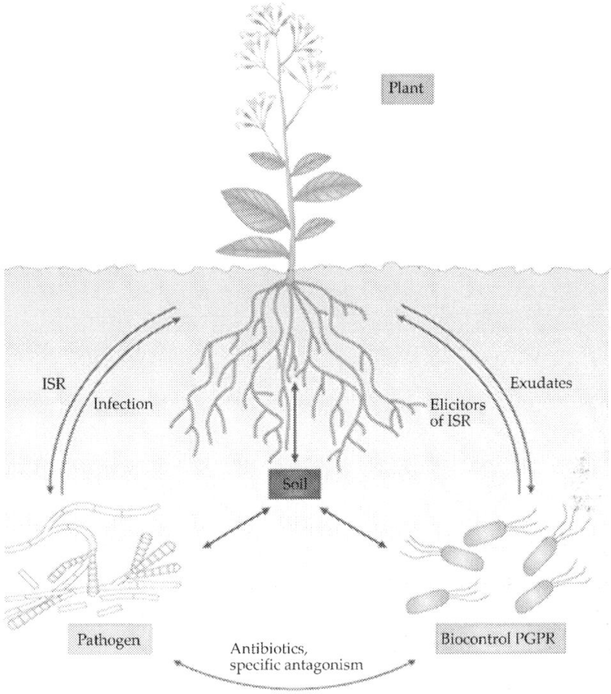

Figure:Interactions between biocontrol plant growth-promoting rhizobacteria (source: http://www.nature.com/nrmicro/journal/v3/n4/fig_tab/nrmicro1129_F1.html)

The direct mechanism of plant growth promotion by PGPR includes increased nitrogen fixation, production of phytohormone and solubilization and uptake of nutrients. In contrast, indirect mechanism of growth promotion includes induced systemic resistance (ISR), antibiotic protection against pathogens, production of HCN, siderophore production, and synthesis of antifungal enzymes or lytic enzymes. The work of Glick and coworkers on PGPR that produce 1-aminocyclopropane-1-carboxylate deaminase (ACC deaminase) and Vanderleyden and coworkers on indole-3-acetic acid (IAA) by *Azospirillum brasilense* in the promotion of plant growth are examples of the mode of action of a PGPR on genetic level.

Solubilization of Phosphates

Phosphorus is an essential macronutrient required in every aspect of cell biology from energy metabolism to the structure of the genetic material, and in soil it is present at levels of 400–1200 mg kg^{-1}. It is the most important key element in the nutrition of plants, next to nitrogen (N), and its cycle in biosphere can be described as open or sedimentary because there is no interchange with the atmosphere. Phosphorus is the most common limiting factor for biological growth and productivity of plants and is taken as phosphate anionsby the plants from soil. Depending upon the quality of the soil, reactive phosphate anions can be trapped by precipitation with cations like Mg^{2+}, Ca^{2+}, Al^{3+} and Fe^{3+}. Usually, a large proportion of phosphorus is found insoluble in soil, but usually a little of total Phosphorus isaccessible to plants which is solubilized by microbes.

The is separate ability of different microbial strains to solubilize unavailable/ insoluble inorganic phosphate compounds such as rock phosphate, hydroxyapatite, dicalcium and tricalcium phosphate (TCP) bythe production of organic acids e.g. gluconic acid, 2-ketogluconic acid, lactic acid, isovaleric acid, isobutyric acid, acetic acid, citric acid, propionic acid, etc. the phosphatases increase plant growth and productivity, thus these bacteria are known as phosphate solubilizing bacteria (PSB). Plants can only absorb Phosphorus in two soluble forms i.e. the monobasic which is phosphoric acid (HPO_4^{2-}) and the other form is diabasic which is a dihydrogen phosphate (H2PO4−) ion form.

Example of phosphate solubilizing bacteria (PSB)

It is reported that a large number of *Bradyrhizobium* strains can solubilize inorganic phosphate. The PSB are common in rhizospheres and they have ability to solubilize phosphates by no means indicating that a rhizospheric bacterium can function as a PGPR, however, it was found in a study that only two of the five rhizospheric isolates positive for Psolubilization actually promote seedling growth. Likewise, not all Phosphorus solubilizes PGPR and increase plant growth by increasing Phosphorus availability to the hosts. The *Bacillus* sp. and*Xanthomonas maltophilia* isolates from Canola (*Brassica napus* L.) rhizosphere has positive effects on plant growth, but no effects on phosphorus content of the host plants.

Biotechnological potentials of phosphate solubilizing microorganisms

Soil has abundance of mineral phosphates, but there is deficiency in the soluble orthophosphate (Pi) which are essential for the growth &development of most plants and microorganisms. Microflora solubilizes the mineral phosphate efficiently including *Bacillus* spp., *Pseudomonas* spp., *Rhizobium, Azotobacter, Azospirillum,* etc. These also play an important role in the supply of nutrients to the plants and therefore help in plant growth thus promoting activities. Numerous reports revealed plant growth promotion by rhizospheric microbes have the ability to solubilize inorganic and organic Phosphorus from soil after their inoculation in soil or plant seeds. There is a mechanism for mineral phosphate solubilization to produce organic acids and acid phosphatases, thus playing a major role in the mineralization of organic phosphorus in soil. Gluconic acid &2-ketogluconic acid are the most frequent agents for solubilization of mineral phosphate. Also, other organic acids such as lactic acid, citric acid &tartaric acid are identified in strains with phosphate solubilizing ability. Therefore the use of PGPR inoculants as biofertilizers provides a promising alternative to chemical fertilizers.

The application of phosphate solubilizing bacterial preparation e.g. the polymyxobacterin in pond aquaculture results in elevated level of mineral phosphate content in water by developing natural forage base which increases fish productivity. In aquatic ecosystems, maximum conditions for growth of *Paenibacilluspolymyxa* were observed. The polymyxobacterin treatment takes part to enhance mineral phosphorus in pond water by 1.3–3.7 times as compared to the control and do not exceed the normative values which are less than 0.5 mg P/dm^3, which also stimulates phytoplankton vegetation and zooplankton development. Due to the development of natural forage organisms, in the nursery and fattening pond the fish productivity increases than the control when after the polymyxobacterin is applied.

Genetics of phosphate solubilizing bacteria

Mainly two components of Phosphorus which are organic and inorganic phosphates are present in soil. Among these two categories a greater per cent of phosphorus is present in insoluble forms and therefore, that

Phosphorus is not accessible for plant nutrition. The important mineral forms of phosphorus are hydroxyapatite, oxyapatite and apatite which constitute the greatest reservoirs of Phosphorus element in soil because under certain appropriate conditions they mineral forms of phosphorus can be solubilized and become available for microorganisms plants.

Genetics of Mineral Phosphate Solubilization (MPS)

The genetic basis of phosphate solubilization is not understoodwell, but the molecular levelstudieshave raised the debatable question of how the PSM solubilizes the insoluble phosphate into soluble form. The production of organic acids from mineral phosphate with the help of enzymes such as acid phosphatases and the gene involved in MPS synthesis have an effect on phosphate solubilization. Phosphate solubilizing gene in Gramnegative bacteria *Erwinia herbicola* are cloned which are involved in MPS by screening the antibiotic-resistant recombinants from genomic library in a medium containing hydroxyapatite as the source of phosphate. The expressionofphosphatesolubilizinggeneallowstheproductionofgluconic acid &mineral phosphate solubilization activity in *E.coli*HB101,thus its sequential analysis revealed its probable involvement in the synthesis of the enzyme pyrroloquinolinequinone (PQQ) synthase which leads to the synthesis of PQQ, a co-factor necessary for the production of the holoenzyme (GDH)-PQQ. Similarly, a mineral phosphate solubilization gene *gabY* isolated from *Pseudomonascepacia*, whose expression gives mineral phosphate solubilization through gluconic acid production in *E. coli* JM109. This induction shows no evident homology with the cloned PQQ synthase gene, but it does show some homology with a permease membrane protein and the napA. The transfer of Phosphatase gene from the soil bacterium *Morganella morganii* to *Burkholderia cepacia* IS-16 occurs which is a strain used as a bio-fertilizer by using the broad-host range vector pRK293. Due to that, the high level of extracellular phosphatase activity of the recombinant strain can be achieved. The information about the genetic &biochemical mechanisms participitates in the synthesis of the GDH-PQQ holoenzyme which is insufficient and alterations between constitutive & inducible phenotypes were reported in several bacterial species, and that has shown that the gluconate, glucose, mannitol, &glycerol are possible inducers of holoenzyme activity of the GDH-PQQ.

However, the manipulation of genes through recombinant DNA technology &molecular biotechnology follows their expression in selected rhizobacterial strainsopen which is an emerging area for obtaining PSM strains with increased phosphate solubilizing capacity and thus, a much effective use of these microbial strains as agricultural inoculants.

Some Future Prospects

the use of PSB are emphasized, as soil Phosphorous resources as an alternative sustainable strategy for the management of soil to optimize Phosphate's bioavailability that has considerable promise for the future and is the newly emerging area for meeting agricultural challenges in varied environmental conditions. PSB gives better opportunity for the development environment-friendly phosphorus bio-fertilizer which can be used as alternatives to chemical fertilizers because the biotechnological and molecular approaches lead to successful plantmicrobe interaction. Hence, further research is needed to find the Phosphate Solubilizing Bacteria (PSB) as bioinoculants in order to enhance the sustainable management of soil & agriculture that helps in economic stability &environment friendliness.

7
CHAPTER

COMPOST BIOFERTILIZERS AS AN ALTERNATIVE TO CHEMICAL FERTILIZERS

COMPOST BIOFERTILIZERS

Compost biofertilizers are those which make use of the animal dung to enrich the soil with useful microorganisms and nutrients. To convert the animals waste into a biofertilizers, the microorganisms like abcteria undergo biological processes and help in breaking down the waste. Cellulytic fungal culture and Azetobacter cultures can be used for the compost biofertilizers.

Biocompost

It is a kind of organic fertilizer, which is prepared from the waste of the sugar industry. The waste is decomposed using a number of human and plants friendly bacteria and Fungi. Biocompost consists of nitrogen, phosphate solubilizing bacteria and plenty of useful fungi like the decomposing fungi. This biofertilizer helps the farmers to increase soil fertility and thereby increase the yield of the crops.

Due to the excessive use of chemical fertilizers, potential pollution of natural resources and Environmental tress are under intensive agricultural systems to meet the demands of the overpopulated world. Then several Health hazards arise due to this system. Therefore, alternative agriculture practices such asorganic farming and use of microorganisms as biofertilizers in agriculture systems is considered a way to preserve the environment and to prevent pollution. Chemical fertilizers suppress the activity of microflora and stability of soil organic matter, In addition to their expensive costs.

In order to increase crop productivity, Organic &biological fertilizers give alternatives to agricultural chemicals which are sustainable ecologically sound in practice. That approach has most recent scope of many researchers. In recent years, the investigations has revealed that the application of organic fertilizers or biofertilizers to soils promote nutrients availability & up taking of nutrients by plants which increase the crop yield also, this reduces inputs of chemical fertilizers thus minimizing environmental risks. There is a biological process in which organic biodegradable waste is converted into hygienic, hums rich product named as Compost. This hums rich product can be used as anorganic fertilizer and a soil conditioner. Against various plant pathogens, the Composing technique can be used in soil to provide biological control. Aqueous extracts of compost are suggested to replace synthetic fungicides. The utilization of compost to agricultural soils gives beneficial effects on yields and crop development by improving biological &physical properties of soil. Not only this, but application of poultry manure increases the herbage, dry matter yield & essential oil content in *Javacitronella* plants efficiently. By microbial interactions in the rhizosphere of plants, the Biological activities are enhanced in plants. Such types of syntrophic associations are of ecological importance showing agricultural significance. (PGPRs) the plant growth promoting rhizobacteria induce plant growth directly by the production of phytohormones and indirectly by nitrogen fixation and production of bio-control agents in response to soil-borne phytopathogens.

Azospirillum* species** are nitrogen-fixing organism named diazotrophs which is capable of forming an associative relationship between the roots of several economically important cereals. According to the Studies, *Azospirillum* promotes plant growth and the exact mechanism of growth has not been fully characterized. Like most organisms, ammonium salt is used as a preferred nitrogen source by ***Azospirillum. The nitrogenase enzyme complex synthesize and convert atmospheric N_2 to NH_4 in the absence of combined nitrogen and under microaerobiosis, *B. cereus*and *B. circulans*are reported as plant growth promoting rhizobacteria and seed inoculation of pigeonpea, wheat and maize with B. cereus, and *Pseudomonas fluorescens.* Although these bio-fertilizers produce beneficial effects on yield and growth in maize had recorded maximum grain yield. Due to seed bacterization with *Bacillus cereus* and *B. circulans,* the increase in grain yield of maize was reported from 43.8 to upto 50.8%, respectively over inoculated control.

Marjoram. is a spice and a medicinal source in the form of the essential oil in aromatherapy. It is used worldwide due to its stimulant and antisasmodic properties. The differences between plant species were reported by the study of chemical composition of marjoram oils. Several workers reported that essential oil components of marjoram wereterpinen-4-ol, gamma-terpinene, trans-sabinene hydrate, linalool, thujanol, terpinolene and thymol.

Due to the ability to produce and store essential oil in perfumes and cosmetics industries, Marjoram cultivation has growing economic impact worldwide. The percentage of fresh and dry matter and essential oils in marjoram plants responded positively to increased levels of composted manure as compared to chemical fertilizer. According to a research, the total nitrogen and organic matter of wasted food compost worked so well with soil microorganisms up to a significant level and increased the fresh weight of lettuce as compared to mineral fertilizers functioning. In addition to such composts, oil production in mint plants was observed to increase when plants were grown with bio-solids. To improve fertility, soil structure, development and productivity of marjoram plants, a little attention has been received to the Application of compost and bio-fertilizers.

Effect of Compost Fertilizer Plant Growth

In a study it was found that plant growth and development of *Bixa orellana* increased with increasing the percentage by volume of FYM added and 30% FYM is recommended to raise standard transplants. In addition, 40 - 50 % FYM is the best to produce transplants in a short time. Fertilized palmarosa oil grass (*Cymbopogon martinii* var. "Motia") by using FYM at 0 and 15 t/ha. FYM increased the mean herb yield by 8%, compared with the control. On *Mentha arvensis* showed that herbage yield was enhanced by the application of the combination of farmyard manure (FYM) and inorganic fertilizer (N, P and K) compared with FYM alone .On the other hand, organically fertilized soil maintained significantly higher available nutrients throughout the crop growth period than non-organically fertilized soil . Stated that the biomass yield of ***Achilleamillefolium*** was increased due to receiving organic fertilizer at the rates of 1 to 4 kg/m2 (cattle manure + straw, which had N,P and K contents of 1.17, 0.21 and 0.95%, respectively).

The addition of organic matter and peat moss increased significantly and progressively the yield of rocket and alfalfa plants grown in sandy soil. Li (1994) found that using cattle manure promoted the growth of *Ocimum basilicum* and *Artemisia dracunculus* plants. on anise (*Pimpinella anisum*) plants, mentioned that the application of farmyard manure (0.46 % N) as anorganic fertilizer in one dose at the time of soil preparation, at two rates (20 and 40 m3/fed.) in the first season, and three rates (10, 20 and 40 m3/fed.) in the second one was less effective (especially at the low rate) than chemical fertilization. Higher N levels had a positive effect on stem length, and raising the rate of chemical or organic fertilization resulted in steady increases in plant weight. The highest rate of organic fertilization (40 m3/fed) increased plant dry weight significantly, compared to the control. Fruit production was more favorably affected by organic fertilization (especially at the high rates of 20 or 40 m3/fed.) than chemical fertilization at low rates.

Figure: Compost fertilizers promote the growth of beneficial bacteria, earthworms, and fungi that build soil structure and foster healthy plants.

COMPOSTING

The one of the oldest techniques used for the stabilization of natural wastes and soil through biological fertilization is termed composting. The main objective of using compost is to obtain a biological and chemical stable and rich product with macro and micro nutrients.

The composting works as follows: Initially, the living waste is broken down into strains of microorganisms. This breakdown generates temperature, while the pH of the medium decreases due to the production of natural acids. the thermophilic bacteria initiate degradation processes, Once the temperature gets close to 40°C and making the temperature to reach 65°C under these conditions the metabolism of certain fungi is deactivated. At this stage, actinomycetes and fungi spore forming bacteriadevelop biological transformation reactions. These degradable compounds such as proteins, sugars, fatand starch are quickly consumed.

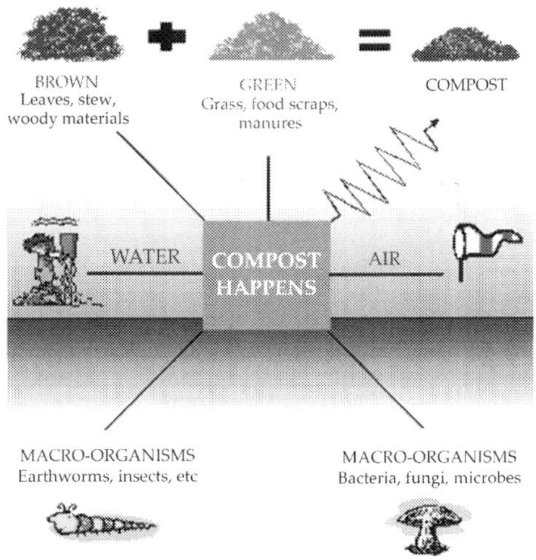

Figure: Process to compost (source: https://www.ndsu.edu/pubweb/chiwonlee/plsc211/student%20papers/articles09/andrea%20fowler/afowler/afowler.html)

In addition to all the other processes happening, due to the release of ammonium ion the pH tends to be alkaline. The temperature and reaction rate decreases, once the organic material degrades. This stage is known as cooling. During this phase, both mesophilic and thermophilic fungi are capable of degrading cellulose. Finally, ripening process begins, which requires several months to be completed. This stage can lead to the complete degradation of compounds and to obtain stable material. The composting process can lead to obtain: stable humus and humic- and fulvicin-acids; characterized by high nutritional value and potential for fertilization of soils with nutriment deficiencies. The compost provides broad Benefits that and can be from the physical, biological andchemical environmental empire. The conditions of organic matter, moisture, temperature, the pH and presence of microorganisms in the pile, depends the application of compost. With structural deficiencies or lack of nutrients, compost increases drainage and absorption of moisture in soils.

The compost permits to

i. increase crop productivity

ii. promote plant growth by incorporation of essential nutrients,

iii. facilitate implementation in different types of soil,

iv. reduce runoff, and

v. Obtain economic benefits for farmers.

Green Manures

To correct and improve chemical and physical characteristics or properties, the green plant tissues incorporates "**Green manures**". Oats, berseem, vetch, rye, clover and peas are Fast-growing crops which are mainly used as green manures. On certain soil characteristics, the use of green manure has a positive influence. For example, soil nutrients susceptible to loss by drainage are retained. On the other hand, certain long-rooted manures have the ability to transport nutrients to the surface and capture nutrients from lower soil horizons, which increases their availability for the development of metabolic processes of plants. The amount of available organic matter in the soil is increased by using Green manures for development of metabolic processes of native flowers and other plant species. Plant materials are susceptible to microbial decomposition, being in direct contact with the soil matrix. This produces humic compounds that are able to increase the adsorption capacity of nutrients, aeration, drainage promotion and soil granulation. In addition, decomposition products serve as a substrate for those microorganisms responsible of biological transformation processes. These processes have a positive impact on the production of ammonia, nitrites, carbon dioxide, nitrates and other simple compounds that are easily assimilated by plants for growth and development.

Finally, to improve soil structure, minimize erosion, and increase water availability in the soil i.e., through evaporation reduction, applications of green manure can be combined with natural inputs. Mineral fertilization with green manure increases yields per hectare and promotes development of mycorrhizal associations. Mineral fertilization with green manure is also a source of essential nutrients and it is a way to foster development of AMF strains.

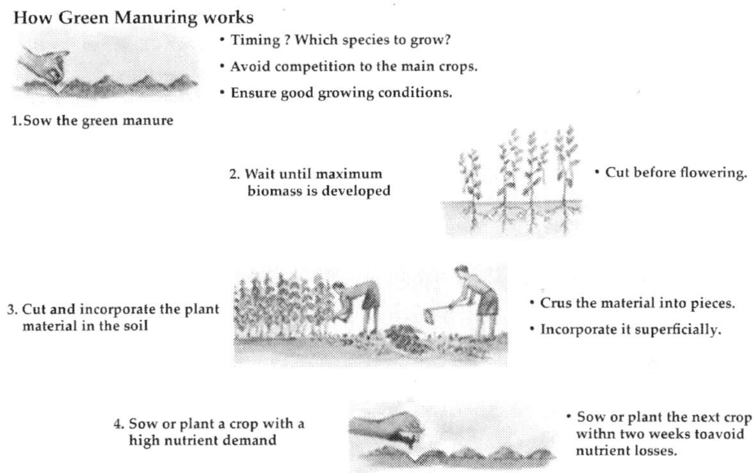

Figure: How green manure works (Source:http://teca.fao.org/read/8368)

Benefits of Compost

The ability to locally produce the pesticides, fertilizers, and other organic ingredients is one of the best features of organic gardening and the easiest among them is compost. Compost is an excellent source of organic matter for garden plants which is agreed by scientists all over the world. Compost provides water, organic matter, air and microorganisms to plants, thus it enhances their growth. It keeps insects, plant diseases, and weeds away from plants, hence it maintains a healthy atmosphere for the soil. The biggest advantage of using compost is that its volume is much lower due to the partial decay of organic matter in compost so it easily becomes part of soil. Furthermore, the microorganism activity increases in the compost, so the concentration of microorganisms becomes very high. For the growing plants, the compost is a concentrated and easy to absorb source of organic matter.

Figure: Compost Fertilizers (source: https://www.ndsu.edu/pubweb/chiwonlee/plsc211/student%20papers/articles09/andrea%20fowler/afowler/afowler.html)

The growers can purchase compost from the market. One can also prepare compost in own garden. Composting on a small scale can be done through the following three techniques:

i. Fast Composting
ii. Slow Composting
iii. Worm Composting

1. **Fast Composting** Fast composting technique is used by many biodynamic farmers. This compost can be prepared in a pit of 1m height, 1m width & 1m length. The pit is filled in with thin layers of leaves, straws & manure. Water is added after adding each layer of soil in between the layers. The pile should be turned regularly to keep it aerated. Compost will be ready within six to eight weeks.

2. **Slow Composting:** Slow composting, is very slow process and that takes so many months for compost formation through this method. Organic matter rich in carbon i.e. brown organic matter such as saw dust is the main ingredient. By filling half of the composting pit with carbon rich organic matter, every day and by adding kitchen waste & vegetable peelings to this composting mixture. The occasionally soil addition along with the kitchen waste. The pit's contents gradually decompose and can be harvesting from the bottom of the pit.

3. **Worm Composting:** Worm composting is known as vermiculture which is carried out in a bin/trench. By filling the bin/trench with soil &organic matter like kitchen waste or vegetable peelings and then adding red worms into the bin. The bin should be moistalways, but too much water should not be mixed. Adding water in large quantities fill pores built by the worms with water thus disturb their colony. A hole can also be made in the bin for purging out any excess water. Some researchers suggest using earth worms instead of red worms, since earth worms are more effective in converting organic matter into manure. In few weeks, the compost becomes ready to use.

Figure: Worm Composting (source: http://teca.fao.org/read/8368)

8 CHAPTER

THE ROLE OF BIOFERTILIZERS IN CROP PRODUCTION

INTRODUCTION

Soil microorganisms play a vital role in the decomposition as well as the availability of plant nutrients, for example Nitrogen, Phosphorous and Sulphur, in order to regulate dynamics of the organic matter. For sustainable agriculture and crops, the microbial inoculants are comprised of essential components of the integrated nutrient management. Other than increasing crop productivity; microbial inoculants can also be utilized as one of the economic input fertilizer dose which can provide more nutrients to be utilized from the soil. Bio-fertilizer as defined is "a substance which is known to help with expansion of the root system as well as better seed germination by the help of microorganism".

A healthy plant always has a healthy rhizosphere generally, which is predominated by useful microbes. On the other hand, in unhealthy soil, the optimum plant growth is not possible in the presence of these predominated pathogenic microbes. Bio-fertilizers are never directly supplying any nutrients to the agricultural crops and they are cultures of specific bacteria and fungi as compared to chemical as well as organic fertilizers. The production technology for biofertilizers is quite simple and their installation cost is also very low as compared to available chemical fertilizer plants.

Microorganisms Used as Biofertilizers and their Functions *Rhizobia*

Rhizobia, are symbiotic bacteria, with the ability fix atmospheric N_2 gas in plant root nodules, have a jointly helpful relationship to their host. The plant roots are entitled to supply freshly synthesized substances along with essential minerals to bacteria. The legumes are less dependent on inorganic N fertilizer as compared to many other non-legume crops due to their N-fixing ability i.e.in cereal crops and pasture grasses. N fixation by legumes also maintains soil fertility as well as they can be of beneficial to majority of crop. Rhizobium inoculation is a renowned agronomic practice which ensures sufficient N supply for legumes in place of the N fertilizer. According to different studies rhizobium can fix about 50-300 kg N/ha. Heavier application of inoculum mixed into peat granules trickled into soil as the seeds are planted is an alternative technique to encourage nodulation.

- Azotobacters
- Phosphate-solubilizing bacteria (PSB)
- Vesicular arbuscular mycorrhiza (VAM)
- Plant growth promoting rhizobacteria (PGPR)

Azotobacters

Azobacters, without any symbiosis, are "free-living atmospheric nitrogen fixing bacteria" in cereal crops. They do not really require any specified host plant. Azotobacters are found abundantly in neutral as well as well-drained soil. They can fix 15-20 kg/ha N per year. *Azospirillum* have the ability to produce antifungal compounds in order to fight against many plant pathogens. They also help in increase germination and energy especially in young plants leading to improved crop stands.

Phosphate-solubilizing bacteria (PSB)

Under acidic and calcareous soil conditions large amounts of phosphorus are fixed in the soil but these nutrients are not available for the plants. Phospho-bacterins are those specialized bacteri and fungi which are attributed to make insoluble phosphorus available to crops. The solubilization of the phosphobacterins is usually due to lower pH of soil and also due to the production of some organic acids that bring about the dissolution of bonds that forms phosphate. Many studies revealed that PSB culture enhances yield up to 200-500 kg/ha and thus 30- 50 kg of the superphosphate is saved.

Vesicular arbuscular mycorrhiza (VAM)

Mycorrhiza is a "mutually beneficial symbiotic relationship" between fungi and the plant roots.VAM fungi is attributed to spread and also infect inside the roots. They are known to possess special structures which are termed as "arbuscules and vesicles". Firstly, the plant roots transmit substances few of these substances are supplied by exudation to the targeted fungi, and secondly the fungi assist in transfering water as well as essential nutrients to the plant roots. The fungal hyphae are able to extend root lengths upto 100 folds. The hyphae when reached into the more wetter and the additional soil areas which help plants absorb many nutrients, predominantly the less obtainable mineral nutrients i.e., zinc, molybdenum, copper and phosphorus etc. Some VAM fungi assume the ability to form a kind of cover around the plant's root which occasionally provides it a cottony as well as hairy appearance. Mycorrhizae have the ability to provide a protective cover against many different problems as for example (i) increase seedling tolerance to drought, (ii) high temperatures, (iii) infections by fungal disease and (iv) even to severe soil acidity. To combat root rotting and soilborne pathogens the application of VAM produces better root systems. The maximum growth response to mycorrhizal fungi is most likely in plants in highly weathered tropical acid soils who are low in basic cations and P, and may also exhibit toxic levels of aluminium. Plants with limited or coarse root systems receive more benifits.

Plant growth promoting rhizobacteria (PGPR)

PGPR correspond to a broad variety of soil bacteria which resulted in stimulation of the host growth, when they are grown in association with a host plant. PGPR modes comprise of (i) fixing N_2, (ii) increasing the nutrients availability in the rhizosphere, (iii) confidently influencing root growth andmorphology and (iv) enhancing other beneficial plant–microbe symbioses. Some latest studies have pointed out that PGPR will frequently have multiple modes of action. A combination of the arbuscular mycorrhizal fungi *Glomus aggregatum*, the PGPR *Bacillus polymyxa* and *Azospirillumbrasilense* maximized biomass and P content of the aromatic grass palmarosa (*Cymbopogonmartinii*) when grown with any of the insoluble inorganic phosphate.

Inoculation of Biofertilizers

Biofertilizers, with or without any carrier for microorganisms are usually applied to seedlings, seeds as well as soils, for example, stickers or composts and peat. Regardless of methodology and protocols adopted, the number of cells who reached the soil from marketable products is fewer than the existing numbers of soil or the rhizosphere microorganisms. These added cells are not attributed to have a beneficial impact on the plant unless multiplication occurs. Additionally, the population of the introduced microorganisms decline and get eliminated in very short time i.e., in days or in weeks. For the success of any biological product few things/parameters are very crucia i.e., the formulation, application procedure as well as storage of inoculum of the product. Smaller shelf life, susceptibility to very high temperature, lack of suitable carrier materials, transportation issues and storage are biofertilizers related issues that need to be solved in order to obtain effective inoculation.

The common ways of inoculation are:

- Seed inoculation
- Soil inoculation

Seed inoculation

Seed inoculation is attributed to the useage of a specific strain of microbe which bears the ability to grow in association with plant roots. The soil conditions must be favorable for the inoculants to give their best. Selected strains of N-fixing Rhizobium bacteria have proved themselves to be very efficient as seed inoculants for legumes. Without antagonistic effects, the seed treatment could be done with two or more bacteria. For seed treatment with Azotobacter, Rhizobium, and Azospirillum along with PSB, the seeds are coated with *Rhizobium* or *Azotobacter* or *Azospirillum* first. Then the PSB inoculant needs to be treated on the outer layer of the seeds. To obtain excellent results, this method provides a greatest number of populations of each of the bacterium.

Soil inoculation

Microbes are to be added directly to the soil in soil inoculation, whereas microbes are required to compete with already present living microbes in the soil that are already very well adapted to local environment and significantly outnumber the inoculum. Inoculants of mixed cultures of beneficial microorganisms have substantial potential for providing a more favorable environment for plant growth and protection and controlling the soil microbiological equilibrium. Therefore, uniformity in performance, sufficient quality control and benefits in all terms must be ensured. Although inoculations with PSBs have not been very successful, joint inoculation of PSBs with corrhizae and N_2 fixing bacteria proved to be successful in co-inoculation of Rhizobium Azospirillum, and combined use of diverse fertilizers.

Combined Use of Chemical and Organic Fertilizers

Due to uninterrupted use of chemical fertilizers, there is increased concerned about their impact on environmental quality. The integrated nutrient management system one of the alternative and is very well characterized by joint use of some chemical fertilizers with the ecofriendly organic materials such as (i) crop residues, (ii) animal manures, (iii) green composts, (iv) manure and (v) reduced input of the chemical fertilizers. However, the management system which fully relies on organic input as plant nutrient sources have diverse dynamics of nutrients accessibility from those which involve the useage of chemical fertilizers. The integrated use of chemical along with organic fertilizer has been demonstrated to be highly beneficial specifically for sustainable agriculture production. Several studies have illustrated the beneficial effect of mutual use of organic fertilizers along with chemical fertilizers to mitigate the deficiency of secondary as well as micronutrients in fields that constantly receive P , N, and K fertilizers without any micronutrient or there organic fertilizer for a few years.

In order to evaluate the impacts of organic and chemical fertility buildup, a field experiment was carried out for about seven years continuously to study several nutrient uptake in mint "*Mentha arvensis*" and mustard "*Brassica juncea*" cropping sequence as well their combined applications were also deeply studied. The results of the study revealed that integrated supply of plant nutrients through FYM farmyard manure and the fertilizer NPK, along with *Sesbania* green manuring, played a vital role in crop productivity as well as on soil fertility. The use of organic fertilizers combined with chemical fertilizers exhibits many positive effects on microbial biomass and hence soil health is also greatly improved as compared to the addition of organic fertilizers alone which was reportedly based on the assessment of soil quality indicators.

In order to increase absorption of K, N, and P in sugarcane leaf tissue in the plant and ratoon crop by the application of organic manure in along with chemical fertilizer as compared to chemical fertilizer alone has been studied and results were significant. Results of the study revealed that all treatments except application of chemical fertilizer improved the soil organic C, total P, N and K status. Enhancement in microbial biomass N and C was prominent in soils receiving organic manures only or receiving the combined application of chemical fertilizers and organic manures as compared to soils which are only receiving chemical fertilizers. The study proved that using both chemical and organic fertilizers is of vital importance in order to maintain soil OM content for balanced fertilization and long-term soil productivity in the tropics where soil OM organic matter content is low.

Combined Use of Biofertilizers with Chemical or Organic Fertilizers

To ensure adequate nutrient supply to the plant, the activity of soil organisms is very vital. They can be very capable of dissolving nutrients and making them accessible to plants if the microorganisms find appropriate conditions for their growth. It was found that the application of PSB, *Bacillusmegatherium* var. *phosphaticum*, increased P availability in the soil and the PSB population in the rhizosphere. It also improves sugarcane quality, yield as well as growth. When used in conjunction with P fertilizers, PSB reduced the required P dosage by 25%. In addition,

50% of costly superphosphate can be altered by a cheap rock phosphate, when applied as mixture with PSB.

The effects of a mixed treatment of multifunctional biofertilizer which is a combination of Bacillus sp. *B. subtillis*, *B erythropolis*, *B. pumilus* and *P. rubiacearum* along with the addition of plus 50% chemical fertilizer on a treatment of with biofertilizer and chemical fertilizer on lettuce growth were compared. The results of the study clearly showed that there was a 25% improvement in lettuce yield for the treatment of ½CF + biofertilizer as compared to that of the chemical fertilizer treatment, this highlighted that at least 50 percent of chemical fertilizer could be saved. As multifunctional biofertilizers were beautifully used along with the chemical fertilizer, the assessed effects of multifunctional biofertilizer on rhizosphere microbial activity and the growth of water celery in a field experiment demonstrated that the dry weight of water celery in the treatment with 50 percent organic compound fertilizer with multifunctional biofertilizer was enhanced by 34 percent in comparison to the treatment with 100% organic compound fertilizer.

Additionally, the beneficial bacterial counts including mineral PSB, cellulolytic bacteria and N_2-fixing bacteria in the rhizosphere of water celery in the treatment were improved 102-104 CFU/ gin.

Table: All about chemical, organic and biofertilizers

Fertilizer Type	Chemical Fertilizer	Organic Fertilizer	Biofertilizer
Organic Matter	None of organic Matter; Directly Supply mineral Nutrients to plants with changing soil conditions.	High organic matter; highly change soil fertility	High organic matter; Highly increase soil fertility
Nuetrient content	Single high nutrient of 64%	All but in low amount	All but low in organics; while high in chemical Biofertilizers
Fertilizer Function	Fast but less than 7 days	Slow for approximately fifteen days	Slow of bio-fertilizer; Fast about chemical biofertilizer

Integrated Use of Chemical, Organic and Biofertilizers

In order to develop an Integrated Plant Nutrition System an increased consideration is now given that sustain and enhances soil productivity through balanced use of all sources of nutrients, including chemical fertilizers, bio-fertilizers as well as organic fertilizers. The basic concept underlying the Integrated Plant Nutrition System is the modification in plant nutrient supply and soil fertility to a most favorable level for supporting desired crop productivity in an incorporated manner through optimization of the benefits from all possible sources of plant nutrients. Excess inputs and surplus bio accumulation of P in soil usually enhance their capability to contribute soluble and particulate P to surface waters and result in P-driven eutrophication. The content of Bray-1 P in the test soil was 69 mg/kg, rated as "High" in terms of P availability. There are many proposals which are available in previous studies that significant rising of Bray-1 P content is not good for a soil with an increases level of P availability from ecological and financial point of view. The Bray-1 P content of the soil was just about twice as high in the Compost-N treatment as compare to the Chemical Fertilizer treatment, presenting significant P bio-accumulation in the soil. This showed that reducing half the quantity of compost and urea combined with inoculants of diverse strains of useful microorganisms has substantial potential to lessen P accumulation in the soil and save the input of chemical and organic fertilizers.

In order to get improvement in both sustainable agricultural production and to defend the surroundings, efficient plant nutrition management should be practiced. Organic, Chemical or microbial fertilizer has both disadvantages and advantages in terms of nutrient supply, crop growth and soil quality. Developing a suitable nutrient management system that integrates use of these three kinds of fertilizers can be a challenge to reach the goal of sustainable agriculture.

9

CHAPTER

LIQUID BIOFERTILIZERS

INTRODUCTION

Liquid bio-fertilizer are the best alternatives to organic and chemical fertilizers which are gradually more available in the market as bio-fertilizer. Beyond so many other benefits which may be achieved from biofertilizers, the main contribution is from population of microorganism's availability through them. These microorganisms create well health rhizosphere and increase the plant growth. The best advantage of a liquid biofertilizer is that no solid carrier is required. Liquid biofertilizers are also developed for prospective application in the field of modem agriculture such as hydroponics and soilless farming systems. Traditionally, liquid biofertilizers are created from fermentation of some effective microorganisms which are suggested to be used within three months. Therefore, the production of long shelf-life and low-cost liquid biofertilizers can be conducted.

Liquid Biofertilizers: Solution to carrier based biofertilizers

There are two basic parameters that define and determine strength of biofertilizers. These parameters are:

- Number of cells

- Efficiency of the microorganisms to fix nitrogen or solubilize phosphates.

Liquid biofertilizers are liquid formulations having the dormant form of desired microorganisms and their nutrients along with the other substances that promote formation of resting spores or cysts for increased shelf life and also provide tolerance to adverse conditions. The dormant form on reaching the soil, germinate to produce fresh batch of active cells. These cells grow and multiply by making use of the carbon source in the soil or from root exudates.

Loads of work has been conducted in last few decades on carrier-based biofertilizers in the perspective of organic food production. Now advanced researches have been started on the production and testing of liquid biofertilizers, in the view of the advantages of liquid biofertilizers over carrier based formulations.

The advantages of liquid biofertilizers over conventional carrier based biofertilizers are give below:

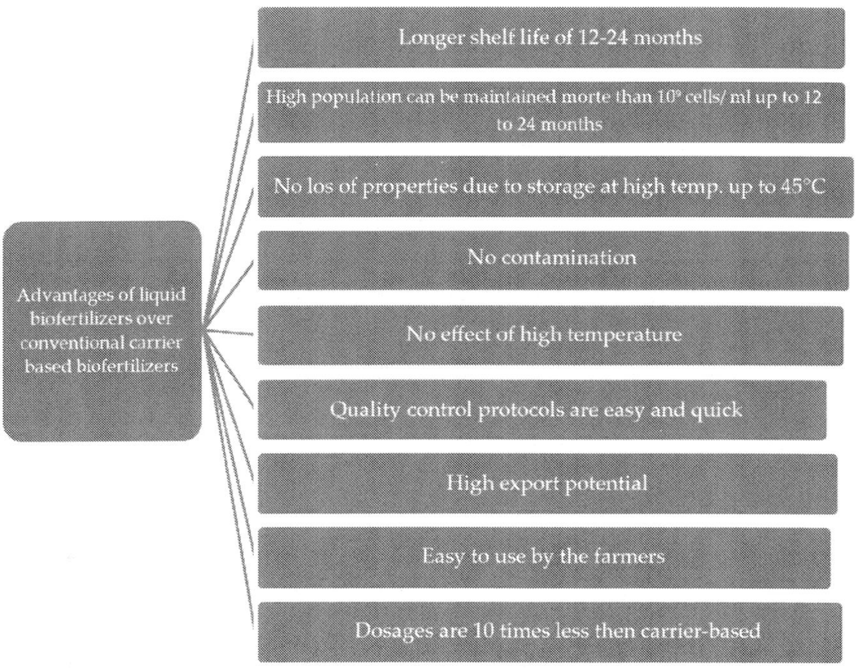

Figure: The advantages of liquid biofertilizers over conventional carrier based biofertilizers

Liquid form of biofertilizer

Liquid biofertilizer is progressively more available in the market as one of the excellent alternative to chemical fertilizer as well as pesticides. The greater benefit of biofertilizer is a contribution from population of available microorganism. Conventionally, liquid biofertilizer acquired from fermentation of effective microorganism (EM) was suggested to be used within three months. Now days the manufacture of ready to use liquid biofertilizer from EM is becoming easily available in market.

Liquid Bio-fertlizer-application/methodology

There are three potential protocols of using Liquid Bio-fertilizers. These are enlisted below:

- Seed treatment
- Root dipping
- Soil application

In the course of biological processes, the biofertilizers acquired the ability to convert nutritional significant elements from unavailable to available from. Some bacteria may fix the atmospheric nitrogen, hence the free living non-symbiotic nitrogen-fixing bacteria from genus *Azotobacter* sp. are heterotrophic and aerobic microorganism which are mostly dispersed in different environments for example water, soil and sediments. Various studies illustrated that due to the improved fixed nitrogen content in soil and due to the microbial secretion of stimulation hormons, the inoculation with *Azotobacter sp.* has valuable effects on plant yield gibberellins, cytokinins and auxins and for the capability in *Azotobacter* sp. to solubilze phosphates. The perfect carriers should have near neutral are greatly adjustable pH, abundant locally at a reasonable cost and able to sterilize. Only a potent and best carrier showed these properties, whereas the final selection for carrier should be based on microbial multiplication and survival during storage, and general method of equipment used for planting, planting and acceptable cost.

Characteristics of *Azotobacter*

Azotobacter, a free–living, aerobic, nitrogen-fixing gram negative bacterium, is also trendy to be used as biofertilizer in order to replace chemical fertilizers. It grows in pH range of 7.0-7.5 while the temperature from 28°C-30°C. For better growth, it uses alcohols, sugar and salts of organic acid. Usually it fixes non–symbiotically about 10mg of atmospheric nitrogen/gm of carbohydrates consumed, usually in the form of glucose. It can form cyst and non-spore formation but in unfavorable circumstances and in older cultures grown with sugar as the carbon source.

Carrier Material

To prepare seed inoculant, a variety of material are used as carrier for seed as well as for soil inoculation. The carrier material is fine powder with particle size of 10- 40μm. According to the well-established and documented studies there are specific properties of a good carrier material for seed inoculations.

These properties are enlisted below:

- Non–toxic to inoculant bacteria strain
- Excellent moisture absorption capacity
- Simple to process and free of any lump-forming material
- Convinient to sterilize by autoclaving or gamma-irradiation
- Obtainable in adequate amount.
- Low in cost
- Good adhesion capabilityto seed.
- Good pH buffering capacity.
- Needless to say.
- Benificial Non–toxic to plant.

Since last 3-5 decades, the chemical pesticides and fertilizers have proved to play key role in increasing the agricultural production. Though, in modern era of agriculture they have a short history. To bring them rapidly in to the center of attention, their immediate action and low cost management are key factors. Their toxic effects on environment, animals, plant and human life deflect the focus on ecofriendly plant protection. Moreover, the development of resistance in insects against widespread pesticides has not been solved yet. Thus, practices such as Integrated Pest Management have gained utmost importance. The Bio-pesticides are vital component of IPM. The terminology 'Bio-pesticide' means "the use of beneficial microorganisms to control the pest". However, considering the total cropped area, the availability of bio-pesticide is the major constraint. Other than the addition of salts to the soil, the differential use of chemical pesticides contribute in loss of soil productivity. A necessary concept of bio-fertilizer which has come forward is to revitalize the soil health and living on alternate source which can be a good replacement for a chemical fertilizers. Bio-fertilizers are nutrient availability systems in which biological process are involved. The term bio-fertilizers include selective micro-organism like bacteria, algae and fungi which are able to convert soluble phosphate and potash in the soil into forms available to the plants, also capable of fixing atmospheric nitrogen. Bio-fertilizers are a renewable cost effective and eco-friendly source of land nutrient. They play a fundamental role in order to maintain a long term soil sustainability and fertility.

The nitrogen fixers as well as phosphate solubilizer along with biofertilizer fixes about 20-40 Kg of nitrogen per acre. The biofertilizer can be produced in liquid form or in solid form as well to be used on the plants. Bio-fertilizer and Biopesticide are the necessity of modern advanced agriculture because demand for residue free & safe food is increasing. However, it is necessary to encourage the efforts for formulation of biopesticides & biofertilizers to furnish the requirement in private sector to encourage the entrepreneurs. Sustainable agricultural is a complex issue associated with producing food while maintaining biophysical resources such as water, soil& biota with no harmful effect on the environment. Biopesticide & biofertilizer are considered necessary for modern sustainable agriculture. Regardless of the fact that biopesticides have been broadly used in developed countries since last few decades, whereas in developing countries the non-availability of the similar products which are suitable for the regionally significant pests resulted in the sluggishness of the growth of biopesticides. Though, in recent are, awareness on biopesticides has been growing with newer products entering the market among farmers of the developing world. And most currently registered products for insect control are formulated in developing countries that include bacteria, nematodes, fungi, viruses and protozoa. As the agrochemical industry adapts to the problems of chemical pesticide resistance, the market for biopesticides is increasing, thus altering public attitudes to pesticide use and re-registration of pesticides has being to happen. In developing countries, similar circumstances are for biofertilizer production. Biofertilizer are available for approximately all crops for three nutrient i.e. K-mobilizers, N-fixers & P-solubilizers. Thus biofertilizer consumption is predictable to increase too many fold in coming years. Since last few years the state agriculture universities as well as non-government organizations were engaged in mass production of biofertilizers. But now several small and medium sector manufacturers have come in picture. And market of biofertilizers.

Liquid formulation of Biofertilizers:

At the moment, biofertilizers are prepared as the most expansively used carrier material, which is lignite and carrier based formulations. The improper ways of handling and improper sterilization of carrier material while packing and combining the bacteria with carrier serve as source of contaminations. Because of this, the inoculant pocket cannot not hold preferred biofertilizer organism for long time, which decrease the shelf-life of the biofertilizers. Further, the quality of the biofertilizer gets worse and due to these unavoided reasons, biofertilizer application cannot give feasible results in the fields. In order to avoid the risks and to enhance the quality as well as shelf-life of bioinoculants, the liquid formulation of phosphobacteria & Azospirillum are formulated. The preservatives can be added with the bacterial cultures and packed in excellent quality bottles. The shelf-life of liquid inoculants is about one year, and they hold the cell count of more than 108 per ml without any contaminants. The liquid inoculants have much enhanced longer shelf-life & superior quality standards without contaminants in comparison to carrier based inoculants.

DETAILS OF TECHNICAL INGREDIENT (ORGANISMS)

Nitrogen fixing organisms

A. Azotobacter (A. chroococcum, A. vinelindii, A. Lipoferum)-Characteristics

Azotobacter, as can fix atmospheric nitrogen into the soil, is a free-living bacterium which is an enormous source to get a natural biofertilizer that can be used in the cultivation of most crops. Azotobacter is a great source of nitrogen in order to meet the needs of crops because it has the potential to cause an alteration of soil microbiology to strike out the biological fixation of nitrogen.

The genus *Azotobacter* consists of bacteria that need the occurrence of oxygen to grow and reproduce, and which are inhabitants of the soil. There are six species of *Azotobacter*. The prominent species is:

Azotobacter vinelandii

This bacterium stain negative in the Gram staining procedure and are of rod-shaped. Some of its species are skilled of directed movement, by means of a flagellum which is located at one end of this bacterium. Furthermore, some species produce pigments to the soil where they are located, lending a yellow-green, red-violet, or brownish-black hue. *Azotobacter* is extremly large, relative to other bacteria. A bacterium is a eucaryotic single-celled microorganism and it can be almost the same size as a yeast cell.

Azotobacter has numerous features that permit it to survive in the sometimes harsh environment of the soil. The bacteria can round up and thicken their cell walls in order to produce cyst. A cyst is either not dormant like a spore but it does permit the bacterium to resist conditions that are otherwise destructive to an actively growing vegetative cell.

B. *Rhizobium (R. Japonicum)*

Rhizobium, a nitrogen fixing, genus of Gram-negative soil bacteria. Along with the roots of legumes, Rhizobium forms an endosymbiotic nitrogen fixing association. Within root nodules, the bacteria colonize plant cells; here the bacteria exchange atmospheric nitrogen to ammonia and then provides organic nitrogenous compounds such as glutamine or ureides to the plant. The plant provides the bacteria organic compounds made by the process of photosynthesis.

C. *Acetobacter xylinum*

Acetobacter bacteria, such as ***Acetobacter diazotrophicus***that can be isolated from sugarcane or coffee plants, are acid-producing and nitrogen-fixing bacteria. In fact, the *A. diazotrophicus* sugarcane relationship, observed and studies in Brazil, was the very first report of a useful symbiotic relationship between grasses & bacteria through nitrogen fixation. In up to date agricultural practices, Nitrogen-fixing bacteria are extremely important as these bacteria reduce the present dependency on nitrogen fertilizers. This characteristic poses positive impacts on ecosystem and on the health of humans and other interacting animals. Other strains can be found in samples from unpolished rice vinegar or Japanese rice vinegar.

Mass Production and Quality Control

A specific medium is formulated in order to grow specific group of microorganisms. During formulation of such medium, one or few additional components are added and /or one or more components are eliminated from the growing media depending upon the necessities of the specific microorganism. Though, according to the physical appearance, media are categorized into two types: a) Liquid media and b) Solid media. The liquid medium is potentially solidified by the addition of solidifying agent agar-agar. Liquid medium can harbor bacterial growth suspended in the media, whereas solid medium harbors microbial growth on the surface. Solid media may be formulated as slant or plate.

Preparation of Starter Culture

The starter culture is a very small amount of bacterial suspension, which is added to the medium in order to start the development of that bacterium. "Twin flask" is a pair of flasks of identical size which are joined together by a latex tube. For the preparation of starter culture, this type of flask is particularly used. Each of such flasks contains a side arm below its neck position. The latex tube joining the two flasks is detained together by this glass tube. The benefit of using these twin flasks is that contamination can possible be avoided.

Fermentation

A fermenter is a device/ instrument in which the most favorable conditions for the microbial growth and activity is customized artificially. This device is used for the formulation of microbial metabolites such as antibiotics or enzymes. Fermenter may also be used for the growth of certain microorganisms i.e. production of microorganism itself.

Sterilization of the fermenter

Fermenter is composed of a metallic vessel for moist sterilization of any targeted material. The principle of moist sterilization lies in the fact that "when water is boiled in a closed system, the water vapor produced due to boiling accumulates within the vessel and increases the inside pressure". Thus the boiling point of water raised beyond 100°C, which is the BP of water in regular atmospheric pressure. In this condition, the steam, produced from the boiling water is of much higher temperature. If any article/ material is kept in this vessel in such condition, the extremely high temperature destroys the microorganisms present in or on the article/ targeted material.

Inoculation, Growth, Quality Testing and Termination of growth:

Inoculation simply means "addition of starter culture to the medium in the fermenter". For the formulation and production of microbial biofertilizers a very little amount of suspension of the targeted bacterium in pure form is inoculated to the experimental medium. Serious care should be given in order to maintain the quality of starter culture, as extent of purity (no contaminants are allowed), size of the starter culture (in terms of culture volume and density of cell) and stage of growth. More the size of starter culture, lesser will be the chance of contamination. If the starter culture is inoculated in its log phase, rapid initial growth will occur. Maintenance of proper physical and chemical environment inside the fermenter is essential for proper growth of microorganism. Quality testing, in this case, is best indication of cell density and its purity in the broth. When the cell density reaches the preferred level, growth is finished and the culture is ready for mixing with carrier. The time period obligatory for optimum cell density is thus standardized.

LABORATORY SETTING AND OPERATIONS

Aseptic techniques

Working in the absence of concerned contaminants is very significant thing. For all pathology work Aseptic technique is indispensable and must be systematically adopted & practiced.

For insect pathology, bacterial cultures or insect pathogenic fungi must be pure. This means that other than the one mandatory, they must be free of any type of living microbes. The presence of unwanted microorganism such as fungi or bacteria is known as "Contamination" and the microbes responsible for contamination are termed as "Contaminants". In order to manipulate or handle microorganisms without appearance of contaminants into the culture, the aseptic technique is used. To protect the operator from possible infection from pathogenic organisms, Aseptic technique helps greatly for this purpose. It is thus recommended to always use aseptic technique when handling microorganisms and also when preparing microbiological media in which to grow these organisms.

Sterilization

Sterilization is "elimination of all viable microbes from a material". Sterilization is one of the nonselective processes. It is very significant stage for any microbiological experimental work. The achievement of proper sterilization ensures the best and required quality of final product. All equipment and media which are required to be used during the handling of the microorganism must be sterilized.

Disinfection

Disinfection is "a way to reduce the contaminant load". It removes potentially infective microbes, but it does not render the object sterile. Many different methods of sterilization can be used. The adopted sterilization method depends on the equipment you have and what it is you are sterilizing. As a general rule, the following methods are most appropriate.

Laboratory growth media

The Sterilization is performed by using dry heat sterilization in an oven or wet heat sterilization for 1-2 hours at 160°C. If there is no autoclave, oven or pressurecooker, then one may use certain chemical agents such as alkalines or strong acids, ethylene oxide or phenols.

Fermentation Glassware

The bio-fertilizer with phosphate & nitrogen fixersolubilizer fixes about 20-40 Kg of nitrogen per acre. The bio-fertilizer preserves the soil fertility cost by using in the yield is assured with bio-fertilizer & consistent use of bio-fertilizer makes the soil very fertile and able to produce good yield of any targeted crop. Application of the microbial bio-fertilizer is a vital step in the Biofertilizer Technology. If the microbial inoculant is not applied properly, it's not possible to obtain targeted benefits from biofertilizers. During application of biofertilizers one should always remember that the most of the microbial biofertilizers are heterotrophic, i.e. they cannot prepare their own food and they greatly depend upon the organic carbon of soil for their energy requirement and growth.

10 CHAPTER

BENEFITS AND LIMITATIONS OF BIOFERTILIZATION IN AGRICULTURAL PRACTICES

INTRODUCTION

Providing of agricultural products, for the food purpose as well as customer goods, the growing concern is to process the current society that has caused considerable increases in agricultural activities in last few decades. In order to get better and efficient crops, reducing in the use of chemical fertilizers is very important. There is need for implementation of improved methods in order to address the issues of concern. This is the reason because of which the implementations of suitable models are a foundation stone of advanced farming practices worldwide. The key focus of conservative agriculture is on limiting undesirable effects on the environment thus increases crop outputs as well as inputs and last but not least implementation sustainable techniques for advanced agricultural developments.

Biological fertilization provides strong foundation for the use of several natural inputs such as (i) decaying remains from organic matter fertilizers, (ii) domestic sewage, (iii) excess crops (iv) microorganisms such as fungi and bacteria and (v) animal manure. They are used in order to get better fixation of nutrients to produce growth stimulants for plants, improve soil stability, for recycling nutrients, to offer better biological controls, to endorse mycorrhiza symbiosis, to biodegrade substances in the rhizosphere as well as in order to develop processes od bioremediation in soils which are poisoned with toxic, xenobiotic and recalcitrant substances.

Moreover, bio-fertilizers can perform different functions when used few of which includes (i) enhance productivity, (ii) consumption of very small portions of energy, (iii) control of contamination from soil as well as water, (iv) increase soil fertility, and (v) promote antagonism and biological control of phytopathogenic. The above mentioned aspects are translated into cost-effective benefits for farmers due to lower costs associated with the process of fertilization and higher crop yields. In this sense, biologic fertilizers application can bring benefits from an economic, social as well as environmental point of view. The implementation of fertilization techniques requires feasibility studies, Proper monitoring of all environment variables of concern which are involved in metabolic processes, acquisition of biological inputs, capital investment, time, and trained personnel.

In a recent research, few experiments were performed in order to assess the implementation of under varying conditions effectively and also to understand the benefits and limitations biological fertilizers as well as to provide a valid background for protocols. The study concluded in the light of results from the use of biological fertilizers in different regions of the world, there is a need for research and focused with a framework in order to facilitate the development of future investigations in the agricultural sector and, thus to promote the reduction of environmental impacts associated with the use of chemical fertilization permanently. Firstly one needs to introduce bio-fertilization techniques before addressing and understanding the fundamentals that explain their functioning. Secondly, numerous benefits as well as the limitations of bio-fertilizers should be followed by for using techniques in various geographical backgrounds.

Biological nitrogen fixation (BNF)

Biological nitrogen fixation is among fundamental ingredient of sustainable agriculture and it is a main important process in the biosphere. It beautifully allows the gaseous nitrogen N_2 to get converted into available form of nitrite, ammonium and nitrate. For the expansion of metabolic processes in plants, the conversion process of gaseous nitrogen and other similar products for more availability for plant's growth takes place by the action of microorganisms in the soil. These microorganisms include: (i) microorganisms that establish associations with grass plants e.g. Azospirillum, Azotobacter, Beijerinckia, Rhizobium, (iii) bacteria that establishes symbiosis with legumes e.g Bradyrhizobium, and Azorhizobium, (iii) symbiotic actinomycetes with woody plants e.g. Frankia , (iv) blue-green algae establishing symbiosis with different plants as for examle Nostoc or (v) ferns as for example Anabahena.

Development of BNF greatly depends on those particular microorganisms who are best carriers of nitrogenase enzyme. Through biological and physicochemical processes they are also responsible for enzyme production. BNF has shown very few to no environmental impacts so far and its usefulness and capability for the optimum physical plant development have been extensively recognized. The gram-negative activity of Rhizobium bacteria is responsible for about 80 percent of fixed nitrogen on the planet approximately. The acquisition approach for reducing atmospheric nitrogen is a complex process by Rhizobium-legume-association. The Rhizobium develops the metabolic cooperation that induces the legume to form nodules; thereby the bacteria reduce nitrogen N_2 to ammonia NH_4. The ammonia NH_4 is transported to the plant tissue to be assimilated into proteins and some other complex nitrogenous compounds. Simultaneously, through photosynthesis the leaves reduce carbon dioxide CO_2 into sugars and transport it to the roots. However, via diatomic nitrogen immobilization, the Rhizobium provides ATP by taking advantage from that source of energy and aided the development of growth and photosynthetic processes of plants.

In addition to the literature discussed in previous paragraphs, it is estimated that Rhizobium-legume association is greatly responsible for setting annually 35 million tons of nitrogen. This amount considerably influences the fertilization of soils globally and supports the development of forestry & agriculture activities in different regions of the world.

Biological Fertilization Techniques

Through agricultural resources and adaptation of modern feasible & effective methods the society must meet its food needs. Therefore the use of these methods in order to meet global demand of inputs and also to obtain better yields has become more and more essential. Similarly, in order to increase the soil fertility there are also few alternative methods and protocols. The main scope of these advanced methods is to provide greater efficiencies as well as to increase the quality of agricultural products to minimize crop production time and to reduce capital costs. On the other hand, due to the continuous and extensive contamination of soils and the use of monoculture and few chemical inputs there is strong need of incorporating few invasive fertilization methods. Many biological fertilization techniques are available and practiced that represent lower environmental, implementation costs and efficiencies comparable to traditional chemical fertilizers used in the world.

Benefits and limitations of biological fertilizers

For the development of better agricultural protocols and practices, biological fertilizers offer a wide range of opportunities due to the advantages and benefits that are providing to the products, soil, and farmers. Nevertheless, very few limitations of adaptation of these practices in agricultural activities are also well recognized, which implies that feasibility studies should be carried out to find out better solutions for each particular case.

Table: Benefits and limitations of biological fertilizers

Benefits	Limitations
Biological fertilizers mobilize nutrients that are favored by the growth of biological activities in soils. The addition of balanced nutrients maintains plant health and also improves its growth. The growth of beneficial soil worms as well as microorganisms is encouraged and food supply is provided. Root growth is greatly increased as a result of the good structure provided to the soil. The content of organic matter in soil is higher as compared to normal levels. Accelerates the development of mycorrhizal associations, which increases the availability of phosphorus P in the soil. Help to eliminate plantar diseases this it provide continuous supply of micronutrients to the soil. Contribute to the maintenance of stable forms of nitrogen (N) and phosphorus (P) concentrations. Improvements on the capacity&exchange of nutrients in the soil.	Their implementation costs are quite higher than those of certain chemical fertilizers. In addition, Compost products have highly variable concentrations of nutrients. They can cause unfavorable effects on human health, plant growth &development as well as deterioration of water quality and death of organisms of the soil is the results of long-term & broad use that results in accumulation of nutrients, salts and heavy metals. For land application large volumes are required in comparison with chemical fertilizers due to very few contents of nutrients. For growth and development of plants, the main macronutrients may not be available in sufficient quantities. Nutritional deficiencies can exist, due to low transfer of macro &micronutrients.

Animal manure

Since last few decades, in order to enhance soil fertilization animal manure has been extensively used by farmers. Their extensive use is attributed to low costs associated with their processing transportation as well as production. This nutritional intake and extensive availability of trace elements make it a smart alternative for the development of fertilized soils that suffer from nutritional deficiencies before.

Figure: Animal Manure (Source: http://www.ecofriend.com/five-reasons-why-you-should-be-buying-organic-meat-and-dairy.html)

- Manure has many potential benefits. Few of which includes:

i. High proportions of potassium (K), nitrogen (N) and medium proportions of phosphorus (P) &calcium (Ca) and low proportion of sulfur (S) and Magnesium (Mg). It allows getting favorable effects on physicochemical stability of plants growth, soils structure and development of beneficial microbial populations.

ii. Manure adds ecofriendly organic matter to the soil.

iii. The composition of organic solids is between 20% - 40%.

iv. Due to high nitrogen content, decomposition of organic matter is also developed more rapidly.

v. Despite having low content of phosphorus (P), manure avoids blockage of this element, making it available for plants.

There are several factors which affects the proportion of macro & micro nutrients available in the manure. These factors include the age, type and health of the animal. For example, poultry & sheep manure contain much high levels of nitrogen (N), while manure from horses & cattle have very lower proportions nitrogen. The type of bedding such as the other plants that serve as disposal of urine and excreta or ferns greatly affects the quality and worth of manure. In addition, the usability & effectiveness of this product depends on the proportion of other chemical substances as well as heavy metals. The application of manure in the soil must be made in concentrations and amounts which are well acceptable by rules and regulations from concerned health & environmental authorities. It is strongly recommended that for large portions of land, the manure should be sprayed as a thin layer rather than stacked on some small portion. By adaptation of this technique its efficiency in agricultural production will be maximized and it will be effective to promote soil aeration and also to facilitate the development of biological activities plants growth which are able to create a medium enriched with nutrients.

Though animal manure promotes plants growth and provides improved and enhanced availability of nutrients, it also exhibits few disadvantages and limitations for example possible risks on the safety of consumers, biological as well as physicochemical stability of soils. In this regard, high contents of ammonia from manure may burn foliage and roots of plants; the presence of manure can sometimes enhance the amount of weed flora and also the costs associated with transportation. The manure application is superior to the conventional and traditional techniques. Other than this, the presence of heavy metals e.g., mercury, lead, chromium etc. pose a significant threat due to their capability of bio-magnification & bio-accumulation in the food chain and their strong carcinogenic potential. For this reason, the evaluation of cost-benefit ratio before using manure to fertilize soils should be well-analysed and well-considered. Also, technical tests must be carried out to verify its safety. Finally, excessive application of manure can also make important reductions of plants growth, extreme levels of nitrogen, ammonia, and salts that could lead to different undesired scenarios for farmers and the soil itself.

Arbuscular Mycorrhiza (AM)

Most plants which are of agricultural interest are endomycorrhiza and they belong to the arbuscular mycorrhiza type (AM). Mycorrhiza is a "mutualistic association existing between most land plants and fungi". These partnerships are quite easy to locate in distinct places, from aquatic to desert, occurring at different altitudes and latitudes. Therefore, its value in terms of accessibility and ease of use in various geographical circumstances is widely documented.

The appropriate selection and application of arbuscular mycorrhiza fungi enhances the resistance of plants against pathogens so enhances plant nutritionand stress including both biotic and abiotic conditions. In addition, the wide range of options and applicability of AM in different regions makes it a faceable technique to replace completely or moderately chemical fertilization.

BioSolids

Dumping of human waste is a common practice since centuries which is treated at specific Sewage Treatment Plants (STP) on the soils. In the early seventies in United States, the application of this sewage sludge in soil or to the sedimentation in agricultural lands and forestry began to obtain beneficial purposes. In fact, the Environmental Protection Agency (USEPA) estimates that about half of the sludge which is produced in the United States is spreaded on the soil to be applied as fertilizer in large portions of land. Bio-solids can also be applied to the soil via injection, dumping as well as by adopting irrigation techniques, depending on the financial and economical conditions as well as local environmental market. These techniques help to decrease (i) insects influence on crops, (ii) spreading of odors, (iii) minimization of runoff and (iv) loss of ammonia in the air.

For using the sludge from STP, several options are appropriate, for example: incineration, landfill disposal and their direct application on the soil. The direct application on the soil necessitate dissolving the sludge, before being applied to the soil, where it is decomposed by the activity of microorganisms and filtered by the soil matrix. Therefore, it is the most promising use from the environmental and economic perspective. Moreover, for soil nutrition, the composition of biosolids is useful. The

USEPA classifies bio-solidsaccording to their content of heavy metals. Bio-solids with higher concentrations are not likely to be used whereas; Bio-solids with lower concentrations can be applied under more flexible security controls. Hence, they must be landfilled or incinerated. The denial or acceptance of bio-solids used as fertilizers is based on safety parameters and hazardous characteristics such as reactivity, corrosively, exclusivity, flammability, toxicity & biological hazards. If bio-solids do not exhibit these characteristics, they can be certified as safe and can be applied to the soil. These sanitary regulations are primarily planned to decrease risks to human health and the environment based on the full potential for contamination of water resources, crops, and ecosystems.

Composting

Composting is one of the very oldest technique which is used for the soil biologic fertilization and also for natural stabilization of wastes. The main aim of this practice is to attain a stable, biological and chemical rich product with macro &micro nutrients.

The process of composting can be explained in following steps: First of all the strains of microorganisms break down the living waste, then they generate temperature differentials while the pH of the medium reduces as a result of the formulation of natural acids. Once the temperature gets closer to 40°C, the thermophilic bacteria initiate degradation processes, thus making the temperature to reach 65°C. Thus under these conditions, the metabolism of specific fungi is deactivated. During this stage, the biological transformation reactions are developed by actinomycetes and fungi spore forming bacteria. These rapidly consume simply degradable compounds such as starch, sugars, fat & proteins. Over and above the processes occurring, the release of ammonium ion results in the pH to get alkaline. The reactions rate lowers as well as temperature, once the organic material is degraded. This stage is said to be cooling. Compost provides benefits which extensive and can be from the biological, physical & chemical environmental areas. Application of compost largely depends on the conditions of moisture, temperature, organic matter, pH &presence of microorganisms in the pile. *For example*, the compost enhances absorption of moisture in soils and drainage with lack of nutrients or structural deficiencies. It also upholds plant growth by incorporation of essential nutrients, it permits to increase crop productivity, reduce runoff to obtain financial benefits for farmers and facilitate implementation in various types of soil.

Figure: Composting (Source: https://www.theodysseyonline.com/panning-black-gold)

Green Manures

Green manures consist of "green plant tissues incorporated on the soil in order to improve its chemical properties or correct physical characteristics". Fast-growing crops such as oats, peas, vetch, rye, berseem andclover, are largely used as green manures. The use of green manure has a positive influence on certain soil characteristics e.g. The Retention of soil nutrients occurs which was susceptible to loss by drainage. On the other hand, certain long-rooted manures have the ability to capture and transport nutrients from lower soil horizons to the surface. Hence, it enhances nutrients availability for the formulation of metabolic processes of plants.

Green manures enhances the amount of available organic matter for the better development of metabolic processes of native flowers and many other plant species in the soil. The plant material is susceptible to microbial decomposition because of direct contact with the soil matrix, which formulates humic compounds that are fully capable to enhance the adsorption capacity of nutrients, aeration, drainage promotion and soil granulation. In addition to this decomposition products serve as a substrate for those microorganisms which responsible of biological transformation processes. These processes pose positive impacts on the production process of ammonia, nitrates, nitrites, carbon dioxide and many other simple compounds which are easily incorporated for development & growth of plants.

In order to enhance and improve soil structure, increase water availability in the soil as well as minimize erosion i.e., trough evaporation reduction, the green manure applications are combined with some natural inputs. For example, mineral fertilization with green manure encourages development of mycorrhizal associations and also increases yields per hectare. It is a way to foster development of AMF strains and it is also a great source of essential nutrients.

Figure: Green Manure (http://agritech.tnau.ac.in/org_farm/orgfarm_green%20 manure.html)

Microbial inoculants (MI)

Microbial inoculants are "biological aggregates or substances containing microbial populations as fermentation bacteria, fungi and lactobacilli". Their elevated nutritional content of salts permits reactions with organic matter in the soil, thus producing favorable substances for plant nutrition e.g., organic acids, vitamins, antioxidants and chelated minerals. Microbial inoculants are capable to modify characteristics of the soil such as macro and micro-flora and can improve biological balance. In addition, their antioxidant properties also promote decomposition of organic matter and improved humus content in the soil matrix. The humus content poses positive and significant effects on quality of harvests, plant growth and improvement of chemical, biological &physical stability of soils.

With a rational use of Microbial inoculants, suppression of biological diseases can be aquired and certain chemical, biological and physical properties can be enhanced. On physical conditions: aggregation of soil particles, improvement of soil structure, reducingsoil compaction and escalating the pore spaces and water infiltration.

- *Chemical Conditions*: improvement of nutrients availability in the soil, leaving free elements to make possible their absorption by the root system.
- *Soil Microbiology*: suppression or control through competition of pathogenic populations of microorganisms which are present on the soil. Microbial inoculants increase microbial biodiversity creating appropriate conditions for the development of helpful microorganisms.

Seaweed Extracts

Seaweeds are comprised of "trace major and minor elements that are helpful for plant nutrition". Other natural substances are also found whose effects are quite similar to those of certain vitamins, growth regulators, proteins, biocidal substances and carbohydrates. They have potential to act against some diseases, pests and chelating agents such as mannitol and organic acids. The significance of seaweed use in agriculture have greater efficiency and better fruit quality which can be obvious from the direct application of these derivatives or pure forms of it. Species such as Ascophyllum nodosum hold micronutrients and macronutrients desirable for cellular nutrition. Few recent studies have showed that vitamin supplements coming from these species can enhance revenues and agricultural productivity. Likewise, it can promote enhanced fruit size, reduce the time of cultivation, availability of sugars, and it can also help to obtain better shapes and tones of agricultural products.

Figure: Seaweeds whose extraxts are used (Source: http://www.agro-chemicals.net/seaweed-extract-flakes.html)

Vermicomposting

Vermi-composting is one of the biological fertilization techniques which is comprised on the use of earthworm`s metabolism in order to produce humus with quite high nutrients content. In order to apply it, organic waste is needed e.g., fruit, peel, manure, crop residues. This organic material passes through the digestive tract of worms, where it is quickly transformed into material rich in specific microorganisms of micro and macronutrients. A biologically as well as chemically stable fertilizer is obtained based on this process.Storage, use, transport, and application of produced vermin compost in soils is of particular interest for those soils which are with nutritional shortages. This technique may be applied as well as developed successfully at both large as well as at small scale in several environments and also under controlled and observation laboratory conditions.

Most widespread earthworm species which is frequently used in vermiculture are Red California and African Red. The Red California presents several advantages which are related to its rapid rate of reproduction under conditions of high temperature of above 40°C, high densities tolerance of 10,000 to 50,000 worm/m^2 as well as resistance to quite large variations in pH, temperature and moisture. Other than these documented facts, under inconsistent conditions, these species are capable to thrive in different substrates. Thus, Eisenia foetida is one of the most used worm in vermiculture. While, on the other hand, Eudrilus eugeniae is also a prolific and rapidly growing worm of concern but it is very hard to manage as the removal from the substrate is quitecomplex as compared to Eisenia foetida.

To get potential benefits from the technique so vermicomposting, to is emphasized in several important studies to take advantage from general climatic conditions as for example from solar-radiation, multi-year average temperature, rainfall, financial conditions for agricultural practices and other social conditions like availability of cheap labor and not only that conditions to incorporate vermicomposting techniques and identify their disadvantages, advantages and limits. A very important and recent study concluded that, the vermicomposting yielded 1.5 tons of humus in tropical zones in one year starting from three tons of organic substrate and 1m^3 of treated soil. Humus resulting from the process exhibited physical, microbiological and chemical properties suitable for application on local soils with nutrients deficiencies such as phosphorus, potassium as well as nitrogen.

Figure: Vermi Composting (source: https://www.indiamart.com/drashana-vermicompost/)

Efficiency and benefits of biological fertilization

Since last two decades, agricultural practices which based on conservative agriculture (CA) have acquired great attention around the world. These have created constructive results on agricultural productivity as well as on sustainability of agriculture, for both conventional and extensive techniques. Many researches focused on biological fertilization have dramatically increased over the last ten - fifteen years. The increasing number of annual publications in this field has been noticed in many renowned journals like Scientia Horticulturae, Crop Production, Food Science and Technology, and Bioresource Technology which clearly indicates an extensive interest of modern researchers on this subject. As an additional remark more professionals and specialized scientists are being required to work on this diversified topic and also to solve different issues around the globe which are about mass production and environmentally friendly practices.

According to a latest research which focused on assessing the influence of green manures (cowpea and Crotolaria sp.), varying doses of cattle manure as well as pea compost over performance of potato crops. Reseals of the study revealed that yields of 55 t/ha (i.e., tones per hectare), 53 t/ha, 47.7 t/ha, 43.3 t/ha, and 33.3t/ha were obtained. These efficient and productive values were not reported before for targeted species in that study area. While, on the other hand, some physicochemical properties of the soil were tremendously improved and environmental impacts (caused by the continuous use of chemical fertilizers) were steadily mitigated.

These used fertilizers may also be consumed to improve yield and performance on tomato crops. These mineral fertilizer saved water up to 25% while 40% pesticide savings and 25 - 30% reduction of mineral fertilizers use were reported in some very important studies. Not only was this increase of 10% yield per hectare observed for crops specifically treated with Arbuscular Mycorrhizal. The resistance of plants to the action of pathogenic microorganisms was quite higher by the use of Arbuscular Mycorrhizal.

As a biological control of fungal species in crops, treatment of cow's feces via aerobic and anaerobic digestion has also been well reported. The scientists had used a bio-fertilizer in order to obtain biological digestion of feces and control the growth of Guignardia citricarpa sp., which, for years, incorporated black spots on the peel of citrus fruits like tangerine, lemon and orange. It was deduced from the results of the study that the compost manure coming from cattle is valuable to obtain best fruits that would be suitable for consumption and sale and for improved biological control of those fruiting plants.

The production of 36% melons, from the incorporation of arbuscular mycorrhizae in crops was also documented in a latest research. Besides that, 100% saving in the use of phosphate fertilizers, 20% saving on potassium fertilization and nitrogen, 25% saving of water, and 100%

reduction in fungicide use was proved and reported from the results of this significant study. In this way, it can be deduced that microbial inoculants (i.e., probiotics) along with polyethylene padding techniques are effective in the cultivation of melons. These technologies can positively control the elimination of pathogenic fungi e.g., Alternaria, Fusarium and Rhizoctonia and can increase proportions of mycorrhizae in the soil.

With the addition of different organic fertilizers, the responses on the photosynthetic process of coffee plants have also been studied and evaluated. The technique of photo acoustic was used in order to compare the photosynthetic activity of several plants treated with bio-fertilizers i.e., from microbial type and chemical fertilizers. The procedure constituted on applying white light, from some artificial xenon lamp, on two 7-months-age coffee plant treatments to evaluate energy storage and the production of oxygen O_2. In order to monitor and assess, pressure detectors as well as photo acoustic were significantly used. The first treatment of chemical fertilizers was received, while the second treatment received was of organic fertilizers. According to this research, organic fertilizers enhances and increases photosynthetic activity, so allowing it to minimizing stress in plants and to grow faster. The conclusion of this research also includes biological products for fertilization of coffee as a sustainable and clean alternative that can be replicated in different geographical areas.

The significant application of compost and bio-fertilizers in oregano crops Majorana hortensis were also observed in a research. A microbial inoculum was shaped along with a combination of Azospirillum brasiliense, Azotobacter chroococcum, Bacillus polymyxa, and B. circulans. Main results of the study indicated that mixed use of bio-fertilizer on crops provides improved physical characteristics of individual plants as compared to those from nitrogen fixing bacteria or compost extracts and also provides better yield performance to the crops. It was summarized that inoculation of oregano (marjoram) with 15% and 30% compost extracts and bio-fertilizer mixtures has significant effects on fat content, dry matter production and plant growth.

The practicability of bio-fertilizers prototypes which is based on native bacteria from rice crops (Oryza sativa) was also studies by a group of scientists. They used in their study a complex mixtures of nitrogen-fixing bacteria with humic substances Polyethylene Glycol, and chelates Carbopol. The production rate of application was considered for a 3-month treatment period. The overall 10% increases in yield production of the mixtures were observed and reported, passing from 7,625 kg/ha to 8,500 kg/ha. The Main outcomes are the importance of bio-fertilizers to get increase productivity and higher revenues, in order to obtain, a sustainable agricultural development progressively.

In order to assess the effects of applications of commercial bio-fertilizers from phosphoric Cerealien and microbial type, in tangerine crops a study was more recently conducted. The research constituted on the application of varying doses of bio-fertilizers solely or mixed in mandarin crops. The response variables took into account was: type and size of fruit and presence of root diseases i.e., rotting of the roots. the use of biological fertilizers, mixed with humic acids, can as increase productivity by 15% as well as reduce the incidence of disease on the roots by 20%, as well and improve physical characteristics of tangerine fruits. The used aquatic cyanobacteria Azolla-Anabaena can be used as fertilizer in rice paddies. This species have been commonly used in rice crops in different regions of the world, especially in Asia e.g., China, Thailand, India and Indonesia. The crop yields achieved closer to 40 kg nitrogen/ha during a 3-month period and verifying increases in growth rate of rice. Furthermore, higher resistance of Milan species to the presence of herbicide Propanil was also found.

A Quantified enhancement in biomass production and growth of Amaranthus and hard pea species by using bio-fertilizers was also well observed i.e., vermicomposting, microbial inoculants, mix of biogas, Azospirillum and Pseudomonas, and a combination of them. Combinedor Sole additions of organic fertilizers were evaluated and the main outcomes in terms of plant growth were that the application of organic fertilizers in combination was much more efficient than individual.

Future Aspects

- The Biological fertilization techniques are valuable and appropriate strategies for effective and efficient use of different agricultural resources with negligible generation of adverse environmental impacts that can pose significant effects on our environment, water resources as well as quality of human life. Other than all these, biological fertilizers also provide a wide range of potentials for the development of sustainable agriculture in different economic, geographic and cultural backgrounds.

- The bio-fertilization tools and techniques require low chemical inputs on soil and helps to facilitate the assimilation of residues that will go to dumping sites and landfills otherwise. The use of bio-fertilizers symbolize applicable reductions on the environmental impacts which are strongly associated to different agricultural activities all over the world.
- In worldwide economies, there is a strong requirement for the recognition of accessible options in order to tackle with the issues and to offer valid frameworks for resultant supply of product for the industry as well as for the development of ecofriendly practices around the globe which allows development and efficiency.
- Feasibility studies must be carried out by at producer end as well as at farmers end. No doubt various options for application of biofertilizers are obtainable to successfully select the best option which offers improved results and minimize environmental impacts.
- In modern agriculture, animal manures, Bio-solids, composting, microbial inoculants, green manures and seaweeds extracts are widely used protocols and techniques, however, their implementation still requires investment, technological developments and research to fully understand their impacts on the human health and soil fauna and flora.

Other Potential Benefits

In order to increase the health and efficiency of plant life, the Biofertilizers exhibit strong potential and also showed potential inorder to reduce the necessity to use synthetic fertilizers. The micro-organisms as well as organic compounds, as opposed to the traditional purpose of fertilizer which is clearly to provide more nutrients, but bio-fertilizers are those compounds that develop strong potential and ability in plants to assimilate nutrients.

Biofertilizer commonly refers to "a broad spectrum of soil amendments and growth stimulators". The common feature in all of these products are that they not only facilitate to cultivate soil fertility in other ways yet they also contain true nutrients, the common example is the Rhizobium bacteria that concentrates nitrogen content in the soil and it self-lives on the roots of specific plants. This fact has already been very well documented extensively by modern science and it also has been recognized since ancient times. From the science of soil biology, the efficiency of other forms of bio-fertilizer is sustained in theory. There are only empirical evidence in terms of soil and plant health and environmental health of their benefits in gardening applications.

- Soil and Plant Health
- Environmental Health

Soil and Plant Health

Majority of bio-fertilizers constitutes of microbes which are involved in the decomposition of minerals and also breakdown of the organic matter into a soluble form which is valuable to plants. Many gardeners may attest to the benefits of introducing compost in the soil while most of the microbes and the compounds which they are creating are abundant in any compost pile. Bio-fertilizers contain particular strains of these naturally occurring organisms that have been cultivated in a lab environment to use for a particular purpose. Some of them are much helpful to fight off diseases and also to provide beneficial trace elements in the soil. While others help in increasing major nutrients levels like nitrogen and phosphorus in the soil.

Environmental Health

Another greatest benefit of using bio-fertilizers is the reduced requirement to use other forms of fertilizer while many of which have negative effects in the environment. For example after longer use, the synthetic nitrogen fertilizers are best known to accumulate salts in the soil thus responsible of making the soil less fertile over time. Either from synthetic or from organic sources, the concentrated applications of nitrogen and phosphates upset the balance of aquatic ecosystems and thus run-off into waterways during heavy rains.

Disadvantages

Despite of the fact that the application of bio-fertilizers did not harm plant life as well as they are safer for environment in any way but still there is little to guarantee that they will help either. The considerable disadvantage compared to nutrient based fertilizers that reliably provide quantifiable results. The reason behind such disadvantage lies in the factor that for the purpose they are prescribed, they have to be associated for the microbes in bio-fertilizers to be effective. The effectiveness of biofertilizers is a product of mixed biological and chemical interactions which are themselves affected by variety of environmental variables just as pH, temperature and moisture. If the conditions are not suitable for the microbes to perform their and multiply, their populations are likely to peter out, and the user have then just wasted time and money on a products which were not suitable for the soil conditions in any way.

Biofertilizers- A key player in sustainable agriculture

Biofertilizers function as an important player in sustainable agriculture by increasing soil fertility, plant's tolerance as well as crop's yield. In order to meet the food demands due to growing human population, the conventional agriculture also plays a noteworthy role. This has led to an increase in dependency on pesticides as well chemical Fertilizers to meet these demands. Chemical fertilizers are industrially manipulated substances comprised of known amounts of nitrogen, potassium and phosphorus. Their utilization resulted in ground water pollution and air pollution due to eutrophication of water bodies. Thus recent efforts have been conducted to ensure bio-safety and more channelized approach towards the formulation of 'nutrient rich high quality food' in sustainable manner. The ground-breaking view of farm production attracts the increasing demands of biologically based organic fertilizers exclusive of the alternative to specific agro-chemicals. In terms of agriculture, alternate means of such soil fertilization are encouraged that relies on organic inputs in order to improve nutrient supply and also conserve the field management. Organic farming is also among such strategies that not only granties food safety but also adds value to the biodiversity of soil. The additional advantages of bio-fertilizers as already described in detail include better shelf life causing no adverse effects to ecosystem.

Organic farming

Organic farming is widely dependent on natural micro-flora of soil which mainly constitutes all kinds of useful fungi as well as bacteria including the arbuscular mycorrhiza fungi called "plant growth promoting rhizobacteria (PGPR)". Biofertilizers maintain the soil environment rich in about all kinds of macro as well as micronutrients via phosphate, potassium and nitrogen fixation. The solubalisation and mineralization helps in release of particular plant growth regulating substances in addition to production of antibiotics along with biodegradation of organic matter in the soil. With the application of bio-fertilizers as inoculants to soils or seeds, they (biofertilizers) multiply and contribute in promoting crop productivity and nutrient carrying capacity. In general 60% - 90% of the total applied fertilizer is lost and the remaining only 10% - 40% is up taken by plants. In this respect, microbial inoculants have great importance in integrated nutrient management systems in order to sustain healthy environment and agricultural productivity. The PGPR or co-inoculants of PGPR and AMF can advance the nutrient use efficiency of fertilizers. A synergistic interaction of PGPR and AMF was improved suited to 70% fertilizer plus AMF and PGPR for P uptake. Same type of trends were also demonstrated in N uptake on a whole-tissue basis which shows that 75%, 80%, or 90% fertilizer plus inoculants were significantly comparable to 100% fertilizer. to cater to the needs of agriculturists and plant biologists whose work focuses on creating efficient and clean means to improve the quality of soil by maintaining and nourishing the natural and useful flora of microorganisms or PGPRs. Further, it presents latest innovations in the area of field management which uncovers the possible application of bio-fertilizers and improved nutrient profiles, plant productivity and growth, and enhanced tolerance towards environmental stress with a specific emphasis on mechanism of the fate of biofertilizers.

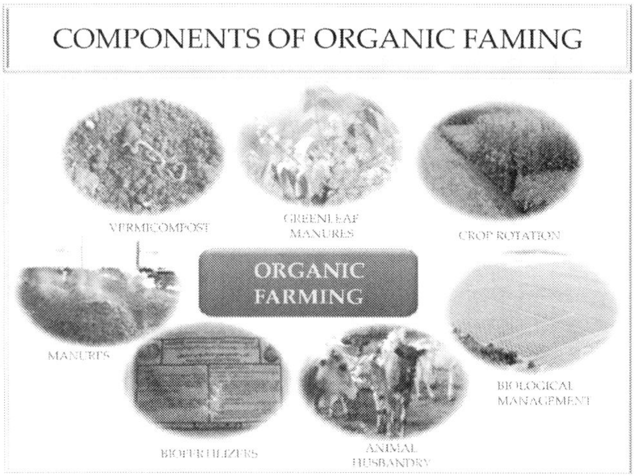

Figure: Components of Organic Farming (Source: UHS Bagalkot, KRCCH Arabhavi, current status of organic production practices in herbal spices, University of Horticultural Sciences, BagalkotKittur Rani Channamma College of Horticulture,Arabhavi, Technology, 2012)

The microbiome: Potential Significance of Beneficial Microbes in Sustainable Agriculture

In the smaller zone of soil which is surrounding plant roots of the rhizosphere, constitutes above 30,000 prokaryotic species in order improve plant productivity and very little microbial cells per gram of root in general. As compared to plants, the collective genome of rhizosphere microbial community which is enveloping plant roots is bigger and is referred to as microbiome. By providing several services to crop plants including nutrient acquisition, organic water absorption, matter decomposition, nutrient recycling, bio-controlandweed control, the Microbiome's interactions concluded about crop health in natural agricultural systems. The metagenomic study supplies the individuals with the core endophytic microbiomes and rhizosphere activity in *Arabidopsis thaliana* using 454 sequencing (Roche) of 16S rRNA gene amplicons. It has been projected that by exploiting tailor-made core microbiome transport remedy in agriculture can be a possible approach in organization and management of plant diseases for different types of crops. Rhizosphere microbial communities provides best alternative for chemical fertilizers which is now subject of great interest in tremendous bio-safety programme as well as sustainable agriculture.

In sustainable production of crops, the main and important point of focus in the future decades will be on application of eco-friendly and much safer methods by exploiting the significant micro-organisms. These microorganisms constituted on diverse characteristics as compared to naturally occurring microbes whose inoculation to soil ecosystem proceeds soil physicochemical properties, soil microbes biodiversity, soil health, crop productivity, plant growth as well as development. The useful microbial populations of agricultural importance cover plant growth promoting, N_2-fixing cyanobacteria, rhizobacteria, mycorrhiza, plant disease stress tolerance endophytes, suppressive beneficial bacteria and bio-degrading microbes. Bio-fertilizers are among necessary and supplimental component towards soil as well as crop management traditions *as for example (i)* organic adjustments, (ii) crop rotation, (iii) tillage maintenance, (iv) residue, (v) recycling of crop, (vi) bio-control, (vii) soil fertility renovation and (viii) of pathogens as well as insect pest. Such operations are be significantly useful in order to maintain the sustainable crop productions. *Azotobacter*, *Azospirillum*, *Rhizobium*, mycorrhizae, cyanobacteria, potassium and phosphorus solubilizing microorganisms are the PGPRs that are found to increase in the soil under minimum tillage or no tillage treatment. Efficient strains of *Azotobacter*, *Azospirillum, Phosphobacter* and *Rhizobacter* can provide significant amount of nitrogen to *Helianthus annus* and to number of leaves, enhances the plant height, seed dry weight and stem diameter percentage of seed filling. Likewise, in rice, addition of *Azotobacter*, *Rhizobium* and *Azospirillum* improves the root morphology and promotes the physiology.

Role of *Azotobacter*

Azotobacter exhibits various metabolic functions as it plays key role in the nitrogen cycle of nature. *Azotobacter* has the potential to produce vitamins i.e., riboflavinandthiamine, and plant hormones i.e. Gibberellins (GA), cytokinins (CK), and indole acetic acid (IAA), and *A. chroococcum* advancing the root architecture, inhibiting pathogenic microorganisms around the root systems of crop plants and improves the plant growth by enhancing seed germination. This genus includes diverse species named as *A. paspali, A. chroococcum, A. beijerinckii* ,

A.vinelandii, A. nigricans, and A. armeniacus. It is also used as one of the best biofertilizer for different crops *few of which includes* rice, wheat, mustard, seasum, linseeds, , castor, sunflower, cotton, sorghum, sugar beets, tobacco and rubber etc.

Role of *Azospirillum*

Azospirillum is motile, free-living aerobic, gram variable bacterium who may flourish in flooded conditions and helps in accelerating several aspects of plant growth as well as plant development, various species of the genus *Azospirillum* includes , *A. irakense A. brasilense, A. halopraeferens, A. amazonense* and *A. lipoferum etc. These species* are documented in various studies to improve productivity of a variety of crops. It was concluded in the results of various studies that *Azospirillum* inoculation have the ability to alter the root morphology by producing plant growth regulating substances via siderophore formulation. It improves and thus enhances root hairs formation in order to provide more root surface area to absorb adequate nutrients and also to enlarge the number of lateral roots. This significantly helps the nutrient profile in the improvement of plant growth and development as well and also improves the water status of plant. Co-inoculation of *Rhizobium meliloti* plus 2,4D and *Azospirillium brasilense* create positive impacts on grain yield and N, K, P content of *Triticum aestivum*. In past eras, *Rhizobium* has been used as an excellent nitrogen fixer. It plays vital role in increasing yield by convertion of atmospheric nitrogen into usable forms. *Rhizobium* is resistant to numerous temperature *ranges and it normally* enters through the root hairs. It forms nodules by uncomplicated multiplying. *Rhizobium* inoculants in different locations and soil types were studied and documented in order to significantly increase the grain yields of lentil, Bengal gram, soybean, pea, and sugar beet rhizosphere and alfalfa, berseem and ground nut. These *Rhizobium* isolates ocquired from wild rice supply nitrogen to the rice plant to encourage growth and development as well. One of the species of *Rhizobium* infects plants other than leguminous plants named as, *Sinorhizobium meliloti* 1021 attributed to promote growth by enhancing photosynthesis performance and endogenous level of plant hormone todeliberate plant tolerance to stress. Rhizobial symbiosis provides defense to plants against herbivoresand pathogens, such as example, green house whitefy *Trialeurodes vaporariorum* and Mexican bean beetle.

Biofertlizers utilization and Nutrients Profile of Crops

The greatest advantage of beneficial microorganisms is absorption and assimilation of phosphorus for their own necessity, which in turn is readily accessible as its soluble form in sufficient quantities in soil. A phosphate-solubilizing bacterial strain *Micrococcus* sp. has polyvalent properties which includes siderophore production as well as phosphate solubilisation. Likewise, in a study two fungi *A. Niger* and*Aspergillus fumigatus* were isolated from decaying cassava peels in order to convert cassava wastes by the semi-solid fermentation technique to phosphate bio-fertilizers. *Burkholderia vietnamiensis is attributed as* stress tolerant bacteria which produce gluconic and 2-ketogluconic acids that are involved in phosphate solubilisation. *Burkholderia* and *Enterobacter* that were isolated from the rhizosphere of sunflower are best known to produce indolic compounds (ICs) and siderophores which may solubilize phosphate. Potassium solubilizing microorganisms are from genus *Aspergillus, Clostridium* and *Bacillus* which are very competent in potassium solubilisation in soil and mobilize in various crops. Mycorrhizal mutualistic symbiosis with plant roots fulfills the plant nutrients demand, leading to enhance plant growth as well as plant development process. This Mycorrhizal mutualistic symbiosis not only protects plants from environmental stress but from pathogens attack as well. By the hyphae from outside to internal cortical mycelia, it leads to the assimilation of phosphate which ultimately transport phosphate to the cortical root cells. Nitrogen fixing cyanobacteria commonly used as biofertilizers are such as *TolypothrixAulosira, Nostoc, Scytonema, Plectonema*and *Anabaena.* Besides the contribution of nitrogen, growth-promoting substances and vitamins liberated by these algae *Cylindrospermum musicola* enhance the root growth as well as yield of rice plants.

Biofertilizers Relevance and Plant Tolerance to Environmental Stress

The major constraints which are affecting the productivity of the crops are biotic and abiotic stresses. Under stress, various tools of modern agricultural science have been expansively applied for crop improvement. In this respect, PGPR's role as bio-protectants has got supreme importance. *Trifolium alexandrinum* inoculated with *Rhizobium trifolii* show enlarged number of nodulation under salinity stress circumstance and greater biomass. *Pseudomonas aeruginosa* has been proved to with stand abiotic as well as biotic stresses and it is also observed that *P. fluorescens* formulates salt-stress induced proteins & osmolytes that helps in overcoming the depressing effects of salt. *P. putida*. Thus with their usage, there is improved germination rate and also improvement in several growth parameters *i.e.* fresh weight, plant height and dry weight of cotton under condition of high salt and alkalinity by increasing the rate of uptake of Ca^{2}, K^{+}, Mg^{2+} as well as by decreasing the absorption of Na^{+}. Few strains of *Pseudomonas* confrere plant tolerance via 2,4-diacetylphloroglucinol. Amazingly, the systemic response was obsereved to be induced against *P. syringae* in *Arabidopsis thaliana* by *P. fluorescens*. Calcisol which is produced by PGPRs *i.e. Bacillus polymyxa, P. alcaligenes*, and *Mycobacterium phlei* provides complete tolerance to salinity stress as well as high temperatures. It has been well documented that inoculations of plant with AM fungi also improve plant growth under salt stress. *Achromobacter piechaudii* also enhances the biomass of pepper plants and tomato under the water stress. Greatly, a root endophytic fungus *Piriformospora indica* is observed to protect host plant against specific salt stress. The inoculation of PGPR alone or along with AM like *Glomus intraradices* or *G. mosseae* results in the enhancement in normal physiological processes in *Lactuca sativa* under stress conditions and better uptake of nutrient. The same plant treated with *P. mendocina* increased shoot biomass under salt stress. Mechanisms involved in osmotic stress tolerance employing microscopic and transcriptomic strategies reveal a considerable change in the transcriptome of *Stenotrophomonas rhizophila* in response to salt stress.

PGPRs as among biological agents which are knownto be one of the best alternatives to chemical fertilizers and they are also known to provide resistance against a variety of pathogen attacks. They may also provide resistance against different pathogens metabolites. Other than acting as growth promoting agents, *Bacillus subtitles* can also induce defense-related pathways namely "jasmonic acid and salicylic acid". Application of PGPR isolates *B. amyloliquefaciens* and *B. pumilus* providing immunity against tomato mottle virus. *B. megaterium* is illustrated from stem of black pepper which acts against *Phytophthor capsici Bacillus subtilis* along with adult composts is found to control *Fusarium* infestation on banana roots. Likewise, *B. subtilis* is known to induce foliar as well as root growth in cotton plants and provide resistance against *R. solani*. In several studies it is proved that along with bacteria and mycorrhizae resistance can be confered against fungal pathogens that inhibit the growth of many root pathogens such as *R. solani, A. obscura, Pythium* spp., *F. Oxysporum and H. annosum* by improving plant nutrients profile &productivity.

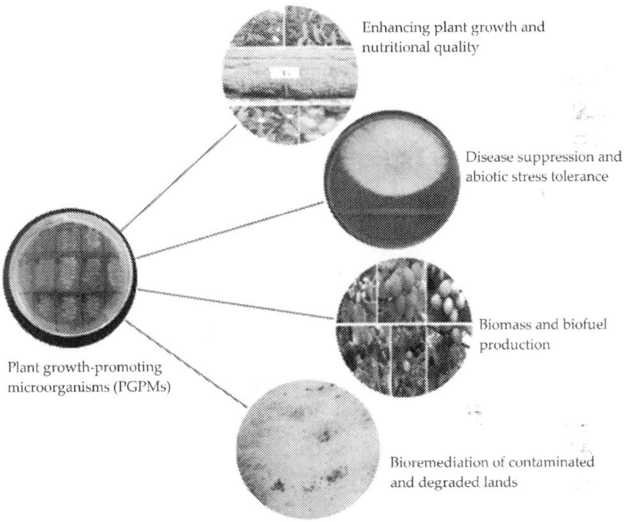

Figure: Plant Growth-Promoting Microorganisms for Environmental Sustainability

(Source: http://www.cell.com/trends/biotechnology/fulltext/S0167-7799(16)30051-8)

Mechanism of Action of Various Bio-fertilizers

Mycorrhiza is the "association of fungus with the roots of higher plants". Mycorrhizal inhabitation act as a model system in order to fully understand the mechanism behind stimulation of growth in particular root cells. Genome sequencing of two fungi ectomycorrhizae, the *L. bicolor* and *T. melanosporum*, helps in the identification of various factors which regulate the development of mycorrhiza and its possible role in the plant cell. Fifteen genes which are known to up-regulate during symbiosis are identified as putative hexose transporters in L. *bicolor.* Its genome lacked genes encoding invertase making it dependent on plants for glucose. However, *melanosporum* possesses one invertase gene which unlike *L. bicolor* can openly be used as the sucrose of the host.

Cysteine-rich proteins of fungus play vital role as effectors as well as facilitators in the production of symbiotic interfaces. During mycorrhizal colonization, many genes related to auxin biosynthesis and root morphogenesis demonstrated up-regulation.

A bacterium like *Bacillus subtilis* may induce 247 genes in cotton plant as compared to control where no PGPR is supplied to the cotton plant. Many disease resistance genes which work via jasmonate ethylene signaling as well as osmotic regulation via proline synthesis genes were differentially expressed with induction.

Among the PGPR species, *Azospirillum* is recommended and known to secrete ethylene gibberellins and auxins. Several plants associated bacteria may also help to induce phytohormone synthesis like lodgepole pine when inoculated with *Paenibacillus polymyxa* had elevated levels of IAA in the roots. *Rhizobium* and *Bacillus* were found to synthesize IAA at different cultural conditions such as in the presence of agro waste as substrate, temperature and pH. Ethylene is responsible for the inhibition of growth of dicot plants. It is observed that PGPR can promote the growth of plant by suppressing the expression of ethylene.

CONCLUSION

At an extraordinary rate, the productivity is waning and the environmental stresses are becoming a serious concern. Our dependency on traditional pesticides and conventional chemical fertilizers has encouraged the flourishing of industries which are producing life-threatening chemicals that are not only hazardous for human consumption but are also attributed to disturn the ecological balance. As agriculture is facing numerous environmental stresses, bio-fertilizers may help to solve the problem of feeding and food security. It is vital to execute its applications in modern agricultural practices and understand the useful features of bio-fertilizers.

The novel technology development in combination with the powerful tool of molecular biotechnology can strongly accelerates the biological pathways of production of phytohormones. These technologies can help in providing relief from environmental stresses. Though, one of the reason that why many useful PGPRs are still beyond the knowledge of ecologists and agriculturists is the lack of awareness regarding improved procedures and protocols of biofertilizer applications to the agricultural crops. The most recent progresses in technologies related to microbial science and genomics and plant-pathogen interactions will definitely help to adopt and optimize the required protocols. The success of the science related to bio-fertilizers greatly depends on their proper application to the field of agriculture and inventions of several innovative strategies which are strongly related to the functions of PGPRs. The greater challenge in this specific area of research lies in the fact that along with the identification of various strains of PGPRs and its properties it is needed to scrutinize the actual mechanism of functioning of PGPRs and evaluation of their efficacy toward utilization for sustainable agriculture for this and future generations.

11

CHAPTER

BIOFUNGICIDES

BACKGROUND

Since last decades, the applications of pesticides including insecticides, herbicides and fungicides have increased noticeably in order to increase crop production by reducing pests that destroy the yield. Pest control by usage commercially available chemical pesticides pose serious damage to crop yield. Thus there is strong need to adopt other possible alternative approaches because the residues of synthetic pesticides which are left in the crops are of serious concern and they pose significant toxic effects to agriculture, animals as well as wildlife like birds, honeybees (and other beneficial insects). These pesticides also pose damaging effects to non-target insects thus disturbing the natural population and balance between predators, prey relationship. The dangerous effects of pesticides are not only restricted to insects and birds but it is also very hazardous to animals and human's health. These pesticides also pollute fresh water by mixing into the ground water source.

The advanced agricultural practices have addressed many problems like prevalence of diseases, harmful pests, droughts and loss of soil fertility. The fungal disease which are putting burden on annual yield production overall everywhere by 20-25% decrease estimating money loss to greater extent.

The use of fungicides reduces crop productivity loss because the chemically synthesized pesticides are no doubt quick in their effect but on the other hand they cause so many environmental and health just as revival of new pests, fungal resistance, loss of soil fertility and persistence of toxic substances when they are consumed by animals and humans. Hence, so much care is required in order to address and handle and use these chemical pesticides and their usage must not exceed the threshold limits. All these problems call for an eco-friendly and sustainable approach inorder to decrease the usage and dependence of chemically synthesized fertilizers and pesticides. The biological controls mechanism is dependent on various features like: competition for survival/ nutrients, the lytic enzymes secretion, toxic metabolites secretion, and direct parasitism are efficient bio-control methods which are best suitable remedies in order to control diseases caused by pathogens. There are so many natural fungicides that can be best alternatives to control fungi and enhancing plant life.

Mode of action

Bio-fungicides are excellent and beneficial fungi and bacteria in colonies form. They attack plant pathogens by upsetting the disease caused by them. They are naturally present in soil that makes them low-cost and easy to find and also eco-friendly and sustainable in nature in comparison to chemical fungicides. Their benefit includes the reduced hazard of pathogens from getting resistant to chemical fungicides when it is incorporated with plants disease management program.

Bio-fungicides control growth of other microorganisms in four different ways:

- Bio-fungicides create direct competition for food or nutrients.
- Bio-fungicides grow a defense barrier around roots or rhizosphere.
- Bio-fungicides shield the roots from damaging attack of fungi.
- Bio-fungicides produce "anti-biosis" which is a toxin for attacking pathogens. Its properties are similar to antibiotic.

Bio-fungicides attack and prey on harmful pathogens.

Working of Bio-fungicides

The bio-fungicides must be applied to plants in a mixed growing media before planting and frequent re-applications to protect new growths and to ensure that enough micro-bacteria are present to control infections in plants. Either the seed is treated or the growing media is treated and making sure that the whole soil volume is treated.

Table: Some Biofungicides: Trade Names and Organisms

TRADE NAME	ORGANISM	DISEASES	REGISTERED CROPS
Actinovate	*Streptomyces lydicus*	Powdery mildew, Downy mildew, *Botrytis, Rhizoctonia, Pythium, Phytophthora*	Most greenhouse ornamentals, vegetable transplants.
Cease	*Bacillus subtilis*QST 713	*Rhizoctonia, Pythium, Phytophthora, Fusarium*	Most greenhouse ornamentals, vegetable transplants.
Companion Liquid	*Bacillus subtilis*GB03	Leaf spots, Powdery mildew, *Botrytis*, bacterial diseases, *Rhizocotonia, Pythium, Phytophthora*	Most greenhouse ornamentals, vegetable transplants.
Contans WG	*Coniothryium-minitans*	*Sclerotiniasclerotiorum, S. minor*	Most greenhouse ornamentals, vegetable transplants & herbs. Soil treatment.
Galltrol	*Agrobacterium radiobacter*K84	*Agrobacterium tumefaciens*	Ornamental nursery stock. Soil treatment.
MycoStop	*Streptomyces griseoviridis*	*Botrytis, Rhizoctonia, Pythium, Phytophthora, Alternaria*	Most greenhouse ornamentals, vegetable transplants.

Plant Shield, Root Shield, T-22 Planter Box	*Trichodermaharzianum*	*Rhizoctonia, Pythium, Fusarium, Cylindrocladium, Thielaviopis*	Most greenhouse ornamentals, vegetable transplants.
Prestop WP	*Gliocladiumcatenulatum* JII446	*Botrytis, Rhizoctonia, Pythium, Phytophthora, Fusarium, Verticillium*	Most greenhouse ornamentals, vegetable transplants.
Regalia	*Reynoutriasachalinensis*	*Botrytis*, Leaf Spots, Powdery Mildew, bacterial diseases, *Rhizoctonia, Pythium, Phytophthora, Fusarium, Verticillium*	Herbs and spices. Soil treatment. Plant health promoter.
SoilGard	*Gliocladiumvirens* GL-21	*Rhizoctonia, Pythium*	Most greenhouse ornamentals, vegetable transplants.

Bio-fungicides do not protect already damaged plant. So it's important to know when to apply bio-fungicides. They are to be applied before the occurrence of any disease. The early application of bio-fungicides helps grow root hairs tremendously and protects from attacking fungi. The bio-fungicides should be applied according to the instructions that are given by the manufacturers just as the chemical fungicides are applied in conjunction with fertilizers and insecticides. Bio-fungicides have short life span than chemical fertilizers and they do not cure diseases but act as barrier to diseases before the onset of infections.

Bio-fungicides are living organisms that are used to protect plants from pathogenic diseases. These are free living fungi or bacteria / actinomycetes which are found over roots, soils and foliar environments. They produce wide range of antibiotics, compete for food and parasitize other fungi. They also bring individual or whole plant resistance to different plant pathogens. Another disease management option is utilizing compost or other growing medium containg micro-organisms to deal with diseases.

Bio-Fungicides functioning:

Biofungicides functions in various ways like:

- Rhizosphere Competence
- Parastitism
- Antibiosis
- Inducing Metabolic Changes
- Plant Growth Promotion
- ***Rhizosphere Competence***: The rhizosphere competence is the most successful strain of bio-control which is the ability to grow in colonies in association with plant roots. They also have ability to colonize whole root for several months and hence compete with plant pathogens for food/ nutrients, space and infection sites. Disease suppression in plants happens due to the competition for glucose in the soil. Bio-fungicide species also metabolize seed and root exudates that normally stimulate pathogen germination or zoospore attraction.
- ***Parastitism***: It is the ability of fungi specie to attack and consume plant pathogens. The Mycoparasitism of biocontrol microorganisms involves directed growth, coiling of hyphae around fungus, contact and binding with host fungus, and also penetration and degradation. Cell wall degradati o n by enzymes is always a part of parasitism.
- ***Antibiosis***: Antibiosis is a process in which fungi or microorganism produce molecules that acts just like antibiotics. These molecules affect other organisms by inhibiting their growth and create toxicity for them.

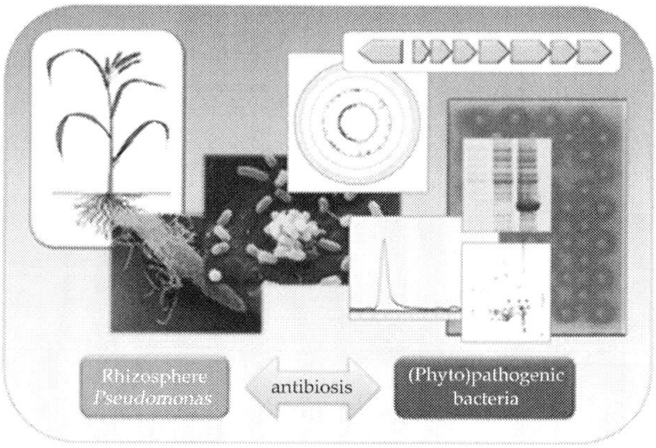

Figure: Biofungicides functioning (Source: https://www.biw.kuleuven.be/m2s/cmpg/research/PGPRB)

i. ***Inducing Metabolic Changes*:** The microorganisms that are used as bio-control has the ability to induce metabolic changes in the host plant or fungi which helps them to increase resistance against many pathogenic diseases caused by bacteria and fungi. There is another defense mechanism by priming the metabolism of plant so that plants can improve their response to pathogens attack. This mechanism is called "Systemic Acquired Resistance". This resistance inducing capacity is prevalent in a variety of plants and is able to cope with wide range of diseases.

ii. ***Plant Growth Promotion***: These microorganisms act as plant growth promoters. The beneficial fungi and bacteria colonize at root which promotes plant growth and productivity. When the plant is not diseased, the bio-controlling bacteria and fungi increase the resistance ability of plants and promote the growth of root and shoot in the absence of invading fungi. Bio-fungicides can be as effective as chemical fungicides when they are applied with growth media such as treated seed for disease control. By studying in detail the characteristics of bio-fungicides it is clear that they are less expensive, more persistent and safe for growers. Bio-fungicides also increase and improve drought resistance in plants when it's applied as seed treatment. Bio-fungicides also increase the uptake of nutrients in plants.

12
CHAPTER

TRICHODERMA

INTRODUCTION

Trichoderma is that fungi that protects plants from molds and bacterial growth. It prevents plant from getting destroyed from root rot and grey mold by creating a barrier that does not let harmful bacteria to invade. Trichoderma has the ability to fight with many soil borne diseases like wilt, rhizome rot, vine wilt and vascular wilt in crops. The chemical fungicides are not only expensive but also hazardous to the environment thus overall making harsh impact. On the other hand, trichoderma is effective and environment friendly than chemical fungicides. But no one uses it in present, apart from its natural abundance. Fungi were used in agriculture dates back to 1930 in all parts of the world.

Figure: Trichoderma

Modes of action

As biofungicides are greatly preventive (rather than curative) thus their application in early the production cycle before the pathogen attacks demonstrated great results. Early application of Biofungicides gives the beneficial microbes. A chance to colonize biofungicides, enables them to out-compete and/or attack the pathogen successfully and thus defending the plant against various diseases. Biological fungicides are quite distinguished from chemical fungicides in their modes of action. Chemical fungicides usually demonstrated two modes of action. The first is by contact, where the chemical product is requiring in order to bring into direct contact with the pathogen. The second mode of action is systemic, where the plant absorbs the chemical. The pathogen is thus adversely affected when it attacks the plant. Since biological fungicides are composed of living organisms, they have significant modes of action that are distinct from those of chemicals. Some biofungicides contain microbes that are antagonistic and parasitic to pathogenic fungi. Quite exactly, the beneficial microbes attack and feed off of the invading fungi. The favorable microbes in other biological fungicides have broad spectrum properties, including the production of antibiotics and/or enzymes. These exudates serious affects and basically kill pathogenic fungi. In their weakened state, the patho-gens are far more susceptible to the affects of chemical fungicides. Many growers find that a two-tiered (chemical/biological) attack on pathogenic fungi leads to fewer applications of chemical fungicides.

Trichodermaharziamum *is common specie of this fungus. It is naturally found native soils and almost in every continent of the world in organic form. To be used to control different fungal pathogens, it works as fungicide and bio-control as it likes to grow in temperature* between 86 to 98.6 degrees Fahrenheit and survive between 77 and 104 degrees Fahrenheit as they die off outside this temperature range.

Trichoderma is very resistant to pathogens, heavy metals and bacteria as they can adjust in any situation. But these fungi are losing their life due to the interruption of human activities and *spoiling their natural existence in all types of soil. Once it was naturally grown but now it is cultivated in colonies as humans destroyed their natural growth by their mining activates extreme climatic conditions, pesticides and aggressive construction works. Thus now we have to add this on in soils rather benefiting from it naturally.*

Trichoderma has many commercial formulations such as T-22 Planter Box, Soil-Guard, Root shield, Pre-stop; plant Shield which has close relation to Gliocladium species. Trichoderma is the most studied bio-control specie of all and is most abundant in numerous composts as they are abundantly isolated from tropical and temperate soils. So the production of antibiotics from trichoderma is well documented. Trichoderma is an active colonizer and strong competitor in toxic environments.

Trichoderma Fungicide for Phytopathogens:

This useful filamentous fungus is very active for various plant pathogens. The plant pathogens are controlled by *Trichoderma spp.* when it makes hydrolytic enzyme which is released to destroy cell wall of plant pathogenic fungi as it controls root diseases in particular when it specifically grows on plant root surface. It must be isolated and characterized as BCA (Biological Control Agent) so that it can be used in integrated pest management. As it is mentioned earlier that Trichoderma is a best control for root rot disease in citrus plants which is caused by *(Armillariamellea).* Since now this natural fungicide is beneficial to control so many soil borne plant diseases *Phytophthora, Rhizoctonia, Sclerotium, Phythium, Fusarium, Sclerotinia,* and *Galumannomyces.*

Presently, entrepreneurs are mass producing its three main species viz, *Trichodermaharzianum, Trichodermaviride* and *Trichodermakoningii* to control multiple diseases.

Trichoderma spreads worldwide and by serial dilution it can be easily isolated as they are known as imperfect fungi and lack sexual stages and faster growth rate in culture media. Trichoderma species are fungi with telomorphs belonging to the Hypocreales order of Ascomycotina division. Trichoderma species are being shown to be the anamorphs of hypocrea species. Hypocrea was first described by Elias Fries in 1825 based on sphaeriarufa pers.: fr., aspecies with hyaline, ascospores.

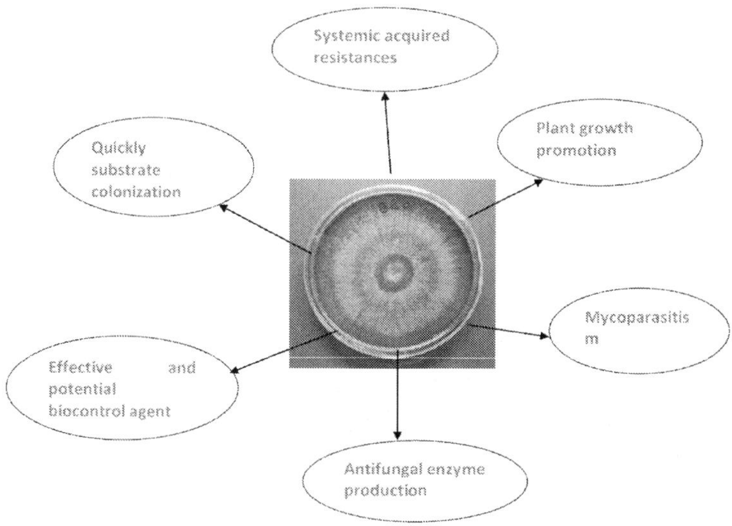

Figure: Biological activities exhibited by Trichoderma

Trichoderma is so beneficial in its working and is very versatile in dealing with food and textile sector on big commercial scale. This is economically good because of the fact that it produces secondary metabolites, hydrolytic enzymes and destruction of xenobiotic and on top of all to develop disease resistance in transgenic plants. Trichoderma is a species which is considered best to study microbial plant ecology, plant protection and biological control. This species is adaptive to many ecosystems like in dessert soils, forests, prairie, pastures and agricultural fields due to its unique feature of growing on any substrate or soils and rapid growth in diverse climatic zones.

Trichoderma grows fast and live longer than other harmful fungi. Their colonies develop fast at the root section thus consuming food and space for themselves and preventing other harmful fungi from growth and multiplication.

Trichoderma is green and it has 1 crore spores in 1 gram of its seed. These spores are taken from soil and are processed in the laboratory by mixing them with talk and lignite powder for 5 to 8 days. After this processing they are stored in cool, dry place away from sunlight.

Applications

It has various applications and uses few of them are:

- Trichoderma for Soil Treatment
- Trichoderma for Seed Treatment
- Trichoderma for Foliar Application

Trichoderma for Soil Treatment

Trichoderma is to be applied with organic matter always like cow dung, sheep or goat droppings, neem powder or compost. To be applied with compost, the 10 kg of trichoderma should be mixed with 1 ton of dried compost. The spores are multiplied within 15 days in a moisture condition of upto 20%. Then this mixture is applied to plants to stop plant diseases to occur.

Figure: Root proliferation of nursery plants due to growth stimulation by Trichoderma applied as soil amendment (left) as compared to non-amended soil (right)

Trichoderma for Seed Treatment

Trichoderma is also used for seed treatment to improve plant's resistance against diseases. The method involves wetting of 1 kg seeds and then adding trichoderma of about 5 to 10 gm. This process follows with thorough mixing of these two in plastic box or sheet followed by sun drying for 2 hours in a shade.

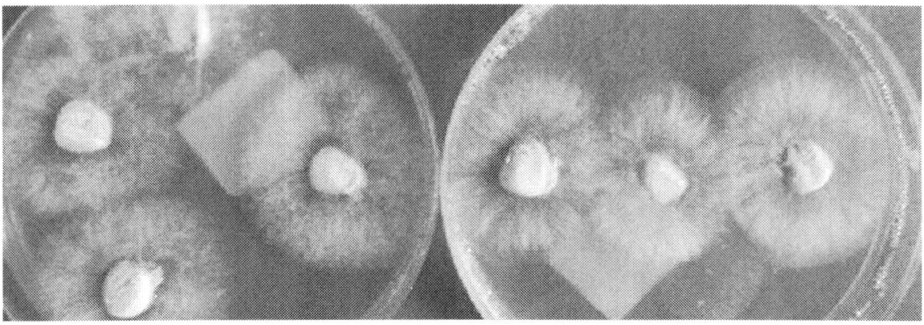

Figure: Trichoderma used for Seed Treatment (source: http://realipm.com/index.php/products_cat/seed-treatment/)

Trichoderma for Foliar Applications

The Trichoderma can be sprayed to plants by mixing 10 g of it with 1litre of water and then sprayed to plants. The use of trichoderma at planting stage is very beneficial because it let the plants grow healthy and to develop immunity against diseases.

Cultivation of Trichoderma

By using liquid and solid state fermentation technique, trichoderma can be evolved into powdered talc form which was first achieved in India by Tamil Nadu Agriculture University.

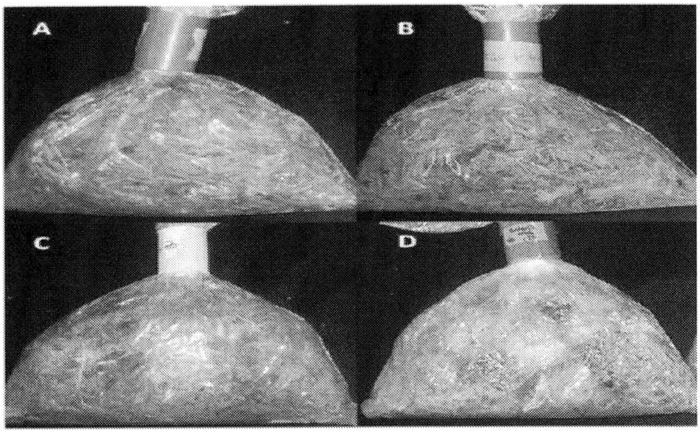

Figure: Growth of Trichoderma Sp in bags A: (-) non-growth, B: (+)Poor Growth, C(++)intermediate growth, D(++++) Best Growth. (source: http://www.scielo.br/scielo.php?script=sci_arttext&pid=S1517-83822014000400017)

The details of solid state fermentation and liquid state fermentations are as follows:

Solid State Fermentation

Trichoderma is commonly formed by this process in which grains of millets, rye, corn and sorghum are used as substrate for its mass production. First of all these grains are dampened, then sterilized and inoculated with Trichoderma Spp. taking it to the next level which is incubation for 10-15 days. Trichoderma covers the whole grain by producing dark green colored spores on them. These grains are further powdered and mixed in sterilized talc powder with ratio of 1:9. This final product is used for the treatment of soil and seeds. This level of Trichoderma mass production is used for commercial scale and is very tiring and time consuming.

Liquid State Fermentation

In this type of fermentation, Trichoderma is taken in shaker in a liquid form and is fermented and incubated. When this process is completed, the fermented media is mixed with sterilized talcum powder in 1:9. In short time, maximum trichoderma can be produced by suitable fermentation conditions like antifoam control, pH, temperature, agitation and aeration and by using appropriate fermentation media the biomass production can be enhanced. The maximum amount of biomass and viable propagules of T. harzinum/ T. viride are achievable in 96 hours of fermentation with good aeration, agitation and temperature control. This information was revealed by the Project Dirctorate of Biological Control (PDBC) Banglore.

Shelf Life of Trichoderma formulation

The quality of a Trichoderma bio-formulation is determined by its shelf life which is the most important factor. The shelf life of trichoderma talc, peat, lignite and kaolin based formulations is 3-4 months. After 120 days, the CFU count of Trichoderma is reduced to 50% in talc formulation. In PDBC study of Banglore, it was deduced that by using various ingredients like (chitin and Glycerol) and heat shock at the end of log phase in fermentation, the shelf life of talc based trichoderma can be increased upto one year.

Application of Trichoderma Formulations

For successful management of plant diseases, there are several methods recommended for biofungicides application. The site of action in Trichoderma plays a very important role in the establishment and delivery of Trichoderma. Seed priming, seedling dip and soil application are the most common methods of Trichoderma application but the rare ones are foliar spray and wound dressing. It has potential applications in

- Seed Treatment
- Seed Biopriming
- Root treatment

Seed Treatment

It is one of the effective and easy methods. Just before sowing the seed it is treated and coated with dry powder or dusts of Trichoderma. The 3-10g per kg seed based on seed size, dry powder of antagonist is used for commercial purposes. T. harzianum, T. virens and T.viride were found to be effective seed protectants against Pythium spp. and Rhizoctoniasolani.

Seed Biopriming

In this method, just prior to the emergence of radical the seeds are treated with trichoderma formulation and incubated under humid and warm conditions. This approach results in quick and uniformed seed coating as compared to simple coating technique as Trichoderma conidia forms layer around bio-primed seeds and germinates on the seed surface and such types of seeds better bear the harsh soil conditions.

Root treatment

Wherever the transplanting is being practiced, the treatment of vegetable crops and rice seedling roots can be done either with dipping of roots in bio-agent suspension of spore or cell suspension of antagonist by soaking the bio-agent in nursery beds before transplantation.

For green house and nurseries ideally, the application of bio-control agents is very much suited before planting or at the time of planting to control numerous soil-borne pathogens.

To control fungal decay, trichoderma is applied as a successful bio-control on the ariel parts of the plant for shrubs and trees scratches.

Precautions

There are certain precautions that need to be followed to have safe trichoderma fungi experience and they are as follows:

- The buyer must carefully observe the manufacturing date and expiry date at the time of purchasing trichoderma as it has shelf life of only 6 months.
- There must be special care taken while transporting fungicide because its seeds die eventually when they are exposed to direct sunlight and heat.
- The chemicals fertilizers must not be used for 20 days after and before the application of trichoderma.

CONCLUDING REMARKS

To control plant pathogens, bio-fungicides are best alternatives to chemicals. The only limitations that are being faced while using them is that lack of sufficient knowledge about the ecology of rhizospere and application of in-vitro antagonism to use biological control agents. But there are several other benefits that these fungicides offer. Trichoderma spp. are the potential bio-control agents investigated commonly found in soil because they can reduce the frequency of diseases caused by pathogenic fungi. There is bacterial specie named Bacillus which consists of Cease and Companion bacteria. There are many bio-fungicides that consist of bacteria that cause soil borne diseases which cause foliar diseases like leaf spots, downy mildew and powdery mildew. There are some bacteria that are not specifically bio-fungicides but they are called "Plant Growth Promoting Rhizo-bacteria" (PGPR) because their growth promoting effects help tolerate diseases in plants. Actinovate, Actino-Iron, and MycoStop are composed of species of Streptomyces, an actinomycetes or filamentous bacteria. They are labeled for both root and foliar diseases. Agrobacterium radiobacter 84 (Galltrol A) is registered for control of Crown Gall (Agrobacterium tumefaciens) and there are also a few Pseudomonas species present in some bio-fungicides (Blight Ban and Bio-Save 10 LP).

As the environmental and commercial requirements for new fungicides become more demanding, the merits of biofungicides over synthetic fungicides become more important than ever. Recently, a break though in biofungicide research was made by semisynthetic approaches using antifungal microbial metabolite as the starting point. As seen in the examples of fenpiclonil, fludioxonil and synthetic derivatives of antibiotic strobilurins such as B-methoxyacrylate azoxystrobin and Kresoxim-methyl, this approach is a promising and effective strategy for the development of new biofungicides with desired chemical and biological characteristics. A good a biofungicide must be versatile. It must be able to be used on a wide range of plants and be effective against a Varity of disease pathogens. A successful product must exhibit some residual (lasting effect) they must be save. Any grower will attest to the fact that each hour spent waiting for a restricted entry interval to end is an hour's worth of valuable time (and potential profit) wasted. In general, U.S, EPA-approved biofungicide have short REIs (four hours or less). The market for biological fungicides is expected to expand due in part to concerns with resistance in existing chemical fungicides, societal concerns about the environment and increasing government regulations/

rest-ricttions. Understanding bio-fungicides and what they can do will enable growers to make informed decisions about how best to incorporate these valuable tools of nature into their production practices.

USEPA Registration

To be legally classified as a biological fungicides, a product must be registered with the U.S. EPA (see the table above). Each product is tested to prove the products' safety to human and animal health and to the environment, as well as its efficacy. Extremely strict quality control must be followed during these testing. This ensures that product contains what it claims and does what it's supposed to.

Table: Few USEPA registered Biofungicides

P Prod-uct	Active Ingredient	Effective against	Application Method	Special storage require	Formulation	Restrict-ed-entry interval	Mode of action	Shelf life	Reap-plication interval	Use with chemical Fungi-cides
compan-ion	Bacillus subtilis GB03	Soil-borne diseases	Drench, irrigation, spray	No	Liquid	4 hours	Antibiotics/broad spectrum proper-ties	2 years+	2-4 weeks	Yes
plant-shield	Trichoderma harzianum T-22	Soil-borne foliar diseases	In medium drench, irrigation	Yes	Granular and wettable powder	0	parasitism	6-12 months if refrigerated	10-12 weeks	No
Mycostop	Streptomyces gris-eoviridis K-61	Soil-borne diseases	Drench, Spray and irrigation	Yes	Powder	4 hours	antibiotic	1 year if un-opened and refrigerated	3-6 weeks	Yes
soilGard	Gliocladium virens GL-21	Soil-borne diseases	In medium (prior to seeding)	Yes	Granular	4 hours	Necrophytic parasitism	1 year if stored in cool dry place	4 weeks	No
Taegro	Bacillus subtilis amyloliquefaciens FZB24	Soil-borne diseases	Drip, drench, not irrigation	Yes	Powder	4 hours	Enzyme produc-tion	1 year shelflife once opened if refrigerated	2-3 weeks	Yes
AQ-10 biofungi-cide	Ampelomyces quisqualis M-10	Powdery mildew	Spray	Yes	Granular	4 hours	Specialized hyper-parasitism	10 months	7-14 days	No

13

CHAPTER

GLIOCLADIUM CATENULATUM J1446 AND CONIOTHRYIUM MINITANS

GLIOCLADIUM CATENULATUM J1446

General

The wettable powder that contains 2×10^8 colony units (cfu) per gram of living spores and mycelium of a specific strain of a natural soil. This fungi Gliocladiumcatenulatum has a moderate control on gluey stem blight called Didymella disease but does control (Mycosphaerella) 'grey mould' (Botrytis sp.) which is 'damping-off' and root diseases caused by Pythium. This fungus does have control over edible and non-edible crops and on strawberries that are grown outdoor.

Gliocladium Catenulatum is a reliable fungus that acts as a tool to control several soil-borne diseases and foliar diseases which contains specific strain of soil fungus. This fungus is suitable for integrated pest management as well as biological pest management programs as it is very effective both organic and inorganic growing culture. The benefits of this fungi consists of that it is safe for humans, environment and to the plants who are directly affected by this. This fungi specie has less risk of resistance and works best in organic production.

Gliocladium Catenulatum" helps reduce chemical fungicides remains in plants. It helps in disease resistance prevention for different typical products. This bio-fungicide controls those diseases that are not controlled by other means. This bio-control also minimizes risks that are often produced by chemical fungicides. This also reduces risk to individuals that are dealing with chemical fertilizers or fungicides by replacing them. This bio-control prevents diseases where there is harvest interval at crops where the chemical fungicides were previously used. This fungi species improves overall yield as it grows on growing media to aid plant growth by colonizing roots.

Gliocladiumcatenulatum J1446 in its active mode creates competition for living space by depriving plant pathogens of nutrition and living spaces as it colonizes plant roots.it breaks down fungal cell walls by secreting enzymes that are harmful for dangerous fungi called Hyper-parasitism.

Gliocladiumcatenulatum strain J1446 suppresses so many fungal diseases on grown vegetables, ornamentals and herbs. The dried cell mass of Gliocladiumcatenulatum J1446 is guaranteed for sale and use by Health Canada's Pest Management Regulatory Agency (PMRA) under the authority of Pest Control Products Regulations and in agreement by Pest Control Products Act.

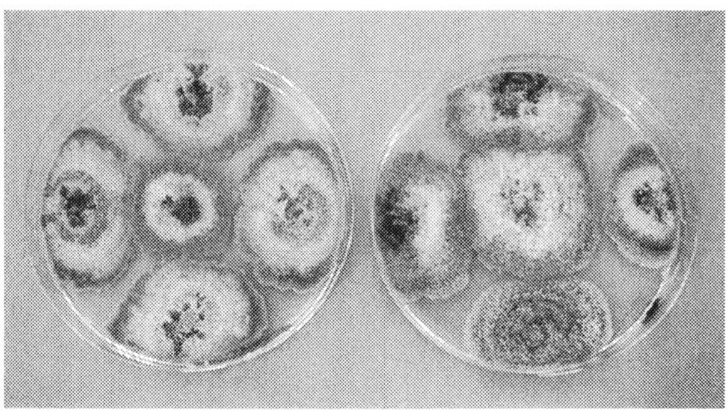

Figure: GliocladiumCatenulatum

Coniothryium minitans

Coniothryium minitans has so many commercial formulations in market. The limitation that this bio-fungicide has that it is less effective when it is applied in spring. The white mold (Sclerotial) population can be reduced by applying Coniothryiumminitans in fall season to the soil before planting.

Cultural Control

There are several management strategies that are effective when combined and are useless when they are applied single to control white mold. There are some practices that are effective to control white mold. Firstly, non-host crops should be rotated for at least 3 years just as sugar beet, corn, sorghum, alfalfa and small grains. The plants should not be planted sin those fields for 3 years where there was a history of white mold before. If plantation is the only option left, then only certified seeds of bush or vine types should be used. Only recommended plant populations should be used and suitable row widths should be adopted for plants growth to facilitate quick drying of soil surface and plants during flowering and irrigation. In order to avoid the canopy formation, the soil type should be tested and according to that only recommended quantity of fertilizer should be used for the purpose. To keep the soil moisture optimal, the irrigation must be scheduled before bloom since disease development is dependent on moist soil so appropriate crop irrigation must be practiced for satisfactory growth. There is only solution to avoid white mold spread in the season is that irrigation must be extended in intervals or its must be ceased without stressing seed fill.

Chemical Control

The chemical controls are most suitable to deal with white mold when they are applied with good cultural practices. In order to protect crops from infection, the recommended quantity of fungicide is to be sprayed during 100% to full bloom. A good protection to plants happens due to good blossoms coverage that falls to the soil surface and in plant canopy.

Role in Controlling White Molds

In disease management, the application of bio-fungicides in controlling Grey and White molds is a new concept. These bio-controls are used differently than registered chemicals. There are some products that require frequent applications of bio-control to give sustained disease control. Three trail applications were made in which disease pressure was low but the result was not encouraging as mold levels exceeded 2% threshold. In order to reduce white mold inoculum levels in soil, the appropriate bio-fungicide is to be used in combination with foliar fungicides over a period of few years. Several other bio-fungicides that should be used as trial to optimize the green-house production system.

Figure: White mold (Source: https://www.royalqueenseeds.com/blog-cannabis-mould-identification-prevention-and-treatment-n198)

Mycofumigant Biofungicides

To control stored vegetables and herbs and to prevent soil borne diseases mycofumigant bio-fungicide was developed. The fumigant activities against phytopathogenic fungus called Botryris Cinerea were tested on healthy tissues of red pepper and trees from the family named Lauraceae. The volatile anti-fungal compounds were produced when two fungi designated as CF016 and EF069 were isolated from cinnamon stems (Cinnamomumloureirii) and red pepper stem (Capsicum annum L). On the basis of morphological characteristics and ITS-5.8S ribosomal DNA sequences study, the CF016 and EF069 were identified as Nodulisporium sp. and Oxyporuslatemarginatus. The O. latemarginatus EF069 which produces volatile gas identified as 5-pentyl-2-furaldehyde which control Rhizoctonia disease on Phalenopsis sp. and grey mold on apples. Under growth chamber conditions, the EF069 suppressed radish damping off by 85% and also the development of Chinese cabbage clubroot retardation upto 33%. The field trials showed that EF069 as very effective bio-fungicide to control clubroot in Chinese cabbage. Hence further studies are required including field trials and large scale fermentation formulations to develop better understanding about O. latemarginatus EF069 producing volatile compounds.

(Reynoutriasachalinensis (REYSA)) Giant knotweed

From hundreds of years plants has been used as cure for several human diseases. But there are few species of fungi that are extracted as fungicides to control plant diseases. Plants extracts are important tool for disease management in organic and conventional systems and further researches are to create to explore more bio-pesticides in this field.

This bio-fungicide is used in many commercial crop productions now a day. This is also used as food in many countries majorly is Asia. It was field trialed through 1990s which was limited to cucumber powdery mildew (*Sphaerothecafuliginea*), Botrytis fruit rot (Botrytis cinerea) and wheat powdery mildew (Blumeriagraminis f. sp. tritici).

The test results of (Reynoutriasachalinensis (REYSA)) Giant knotweed for powdery mildew of cucurbits, downy mildew of lettuce (Bremialactucae), Botrytis of grapes and strawberries, bacterial spot of tomatoes and peppers (Xanthomonascampestrispv. vesicatoria), Cercospora on soybeans (Cercosporakikuchii) and bacterial canker on citrus (Xanthomonasaxonopodispv. citri) applied as foliar spray to control the above mentioned fungal and bacterial diseases. This bio-control induces resistance in plants and thus deals with wide spectrum of disease control. It also induces the production and accumulation of phytoalexins in plants and it also increases chalcone synthesis and chalconeisomerase in phenylpropanoid pathway. Some simple phynolic compounds also accumulate which are fungitoxic in nature.

This fungicide increases papillae formation at sites where pathogens penetrate as well as lignification of cell wall in plants increases according to the studies. The pathogenesis –related protein activity also rises which includes chitinase, peroxidase and glucanase. This fungi act in synergistically with other common fungicides such as azoles, strobilurins and sulphur to control diseases like powdery mildew, copper control in bacterial diseases and leaf spot disease. The Downey mildew is also controlled in mancozeb and mefenoxam thus making it an excellent tool for fungicide resistance management.

There are multiple disease control methods to control soil borne diseases such as seed treatment, irrigation application, dipping seeds before transplantation and soil drenches. This fungi is proved successful due to its unique formulations, diverse application options, through excellent marketing and extensive research based formulations. The human health and environment can be protected if the plants extracted materials are used to increase crop yield, low synthetic fungicides residues on plants and reduced pathogens resistance in plants and soil.

There are certain plant species named Reynoutriasachalinensis, Fallopiasachalinensis) and (Syn. Polygonumsachalinensis of plant family polygonaceae which are known to be antimicrobial due to the fact they contain anthraquinones like emodin, physicon and resveratrol. This fungus has good effectiveness and broad spectrum of dealing with diseases. Many researches have been done which conclude that R. sachalinensis extracts induce plant resistance by increasing defense related proteins production, by phenolic compounds production and phytoalexins, papilla formation and lignification in cell walls and reactive oxygen species accumulation.

14

CHAPTER

BACILLUS SUBTILIS & STREPTOMYCES

BACILLUS SUBTILIS

This saprophytic bacterium has natural strains which are commercially available. It produces anti-biotics, interferes with attachment to the plants and displaces pathogens by inhibiting spore germination. It induces SAR and acstivates natural defense mechanism thus improving plant immunity against bacterial pathogens. Under certain conditions, growers sometimes use copper fungicide rather than using Bacillus to combat bacterial diseases and to reduce phytotoxicity, which happens due to repeated and excessive copper sprays. Fungal leaf spots can be cured by Bacillus as it colonizes on plant rhizosphere and deplete harmful bacterial pathogens and fungi thus stimulating plant growth.

Blight disease caused by fungi called BotrysisCinerea is a very dangerous and common disease of crops causing huge economic losses. Bacillus subtilis Strain QST 713 is an active ingredient of biological fungicide Cease which controls plant fungal pathogens.

The sporulation of B. cinerea can be reduced by using Cease and Veranda O in alternate combinations on weekly basis. These fungicides alone do not give significant results but using them in combination is a best practice. Their modes of actions were studied by spraying them on plants once per week for four weeks followed by inoculation of spore suspension four hours later and bagged to maintain sufficient humidity.

These both fungicides gave high disease controls than other treatments. The rotational use of bio-fungicides can be best alternative than using conventional fungicides as Cease and Veranda gave good disease control to fungicides when applied on weekly basis to treat “botrytis gray mold”.

Streptomyces

It is a naturally occurring bacterium found in soil; it is filamentous and grows on decaying vegetation and soil which produces spores and antibiotics. Streptomycin is basically derived from Streptomyces grisevirichis K 61 isolated from S. lydicus WYEC 108 and sphagnum peat found naturally in soil.

Root-Shield

Pythium spp. is a serious disease of geraniums causing root rot in a variety of crops. To deal with this disease, Root Shied is a good controlling fungicide having an active ingredient named TrichodermaharziannumRifai strain KRL-AG2. It attacks other parasites and fungi with different mechanism and grows naturally in different soil habitats. Trichoderma. spp colonizes plant roots and develops on roots.

There has been limited work done on Root Shield fungicide and on its working ability to control pythium root rot of geraniums. Root Shield’s working ability on two different pythium species were tested and the results showed that Root Shiled controls pythiumdpending upon the environment and inoculum load.

Cure For Brown Rot Disease

Brown rot is a most common spread disease of stone fruits caused by Monilinia. Spp which is the reason of economic loses of pre and post-harvest. The three Monilinia spp. are responsible for brown rot disease namely, M. ftuctigena, M. fructicola and M. laxa. There are many factor that increase the incidence of brown rot in stone fruits such as moisture, wind and insect vector’s movement.

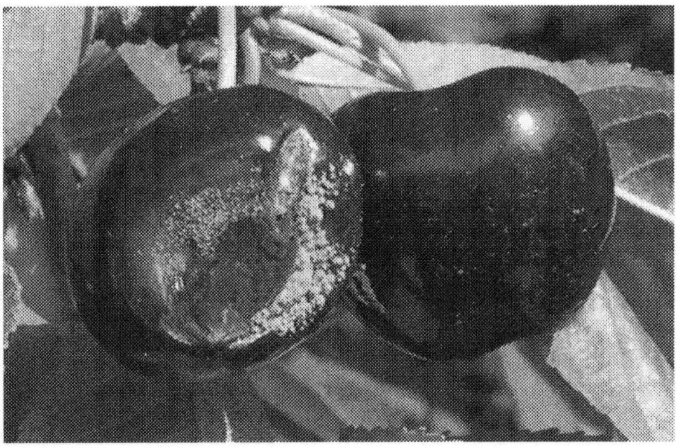

Figure: Brown rot disease (source: http://www.pinsdaddy.com/brown-rot-cherry-tree-diseases_BS6GB5BSDfNeHSv2Y0WetKm1uiSajaZ8wLcoiq9vJX4/)

The brown rot disease incidence can be decreased by following some precautionary measures like fungicides and insecticides should be applied at proper intervals of before and after harvesting. The fruits should be handled with care to avoid the wounding and should be kept in 0°C temperature making the harvested fruit rapidly cooled along with good sanitation.

Brown rot disease can also occur during post-harvest intervals and can cause much crop loss. Sometimes the flowers get infected from fruit rot during post-harvest, harvest and pre-harvest intervals. At any development stage, the fruits can get infected with Monillinia spp. and the disease gets worse when the fruit starts too ripe.

When the pericarp and endocarp of peach fruit is completely formed the M.laxa infection susceptibility increases between 32nd and 34th week of growth. The fruit loss is much during the post-harvest then pre-harvest phase and fruit storage and transport are the other reasons of loss.

The brown rot disease can be controlled by synthetic pesticides but plants build resistance against them with time and too much exposure. The synthetic fungicides are undesirable to control brown rot as there is several health risks associated with synthetic fungicides and insecticides so the public strongly refuses their use as controller for diseases.

Micro-Organisms Relationship

Microorganisms are functionally and nutritionally interconnected and changes in any form do affect the whole community. This happens mostly in plant microorganisms which are influenced by plant metabolic status.

Fungicides have negative impacts that are hard to predict as their mode of action impacts membranes, nucleic acids, respiration, cell division and mitosis, multi-site activity, protein synthesis and signal transduction.

Prevention for New Plugs

There are remedies for plugs that arrive on plants but do not grow for weeks. Vinca is a summer bedding commercial crop just like pansies. It develops chlorotic leaves, poor flowering, slow growth and brownish tan clubby leaves which resulted in poor crop quality and late shipping to landscapers. There is a disease called "Baker's Dozen" which causes same identical symptoms. The secret to grow quality vincais is to keep root environment optimal. There are some solutions to Vinca crop problems.

- Poor crop timings / early planting
- To grow strong roots, high intensity light and warm environmental conditions is required.
- To keep the greenhouse warm, the heating temperature of 70`C is expensive.
- Vinca is not supposed to grow in the months of January and February.

REFERENCES

1. Mishra D.J, Rajvir S, Mishra U.K, and Kumar S.S, "Role of Bio-Fertilizer in Organic Agriculture", A Review, *Research Journal of Recent Sciences*, 2(2277-2502) 39-41 (2013).
2. Beneduzi A, Ambrosini A, and Passaglia, "Plant growth-promoting rhizobacteria" (PGPR): Their potential as antagonists and biocontrol agents, *Genetics and Molecular Biology*, 35(4) 1044-1051(2012).
3. Tjamos E.C,Tjamos S.E, and Antoniou P.P, "biological management of plant diseases: highlights on research and application", *Journal of Plant Pathology,* 92 (4) 17-21(2010).
4. Youssef M.M, Eissa M.F.M, "Biofertilizers and their role in management of plantparasitic nematodes"- A review, Journal of Biotechnology and Pharmaceutical Research, 5(1) 001-006 (2014).
5. Himachal Motghare and Rashmi Gauraha O/o Dy.Director of Agriculture, Raipur (2012). Biofertilizers - Types & their application. http://www.krishisewa.com/articles/organic-agriculture/115-biofertilizers.html
6. Balaram M, Deepak K V, Anindita S, Bipin, Bihari P, Bavita A, "Popular Kheti", 1(4) 1-9 (2014). Article: http://www.biotecharticles.com/Agriculture-Article/Biofertilizers-Types-Benefits-and-Applications-172.html
7. M Mohd, A.Taqi, "Future of Bio-fertilizers in Indian Agriculture",*International Journal of Agricultural and Food*

*Research,*255(2) 571-586 (2014).

8. Fauzia Y, Hafeez, S Yasmin, D Ariani, Y Zafar, K A. Malik, "Plant growth-promoting bacteria as biofertilizer",*Agronomy for Sustainable Development*, 26(2) 143–150 (2006).
9. R. Hilda , F.Reynaldo ,Phosphate solubilizing bacteria and their role in plant growth promotion,*Journal of biotechnology advances,*17(4-5) 319-339(1999).
10. S. Chien, C Young, C Wang, "Current Status Of Bio-Fertilizers Development, Farmers Acceptance And Utilization, And Future Perspective In Taiwan", *Journal of Plant and Soil*, 255(2) 571–586 (2003).
11. M.Mazid, A.yaqi, "Future of Bio-fertilizers in Indian Agriculture",*International Journal of Agricultural and Food Research,*2(3) 10-23 (2014).
12. V. Selvin, J Normon, S Paramakalyani , "Adoptation,problems and future of biofertilizer Ecology", *Environment and Conservation Paper*, 7(**4**) 471-475 (2001).
13. P. K. Sahu , G. P. Brahmaprakash, "Formulations of Biofertilizers – Approaches and Advances", 4(5) 179-198(2016).
14. D Duvauchelle, Natural Resource Specialist, Hoolehua Plant Materials Center Hoolehua, Hawaii, Understanding Inoculants, USDA-*Natural Resources Conservation Service* Pacific Islands Area (2014).
15. L Herrmann, DLesueur, "Challenges of formulation and quality of biofertilizers for successful inoculation", *Appl Microbiol Biotechnol* 97(7), 8859–8873 (2013).
16. D Bhardwaj, MWAnsari, R K Sahoo, NTuteja, "Biofertilizers Function As Key Player In Sustainable Agriculture By Improving Soil Fertility", *Plant Tolerance And Crop ProductivityMicrobial Cell Factories 56(1),* 13-66(2014).
17. Malusa E, Sas-Paszt L, Ciesielskal J, "Technologies for Beneficial Microorganisms Inocula Used as Biofertilizers", *The Scientific World Journal Volume, (*2012).
18. Gupta S, Dikshit AS,"Biopesticides: An Eco-Friendly Approach For Pest Control", *Journal of Biopesticides,* 3(1), 186 – 188,(2010).

19. Rao AS, "Organic Farming: Its Relevance ToThe Indian Context", *Current science,* 88(4), (2005).

20. Paul N , Cruz PC , Aguilar EA, Badayos RB, and Hafelel S, "Evaluation Of Bio-fertilizers In Cultured Rice", *Biofertilizers & Biopesticides ,* 4(1), (2012).

21. Brar SK, Sarma SJ, Chaabouni E, "Shelf-life of Biofertilizers: An Accord between Formulations and Genetics", *Biofertilizers & Biopesticides* , 3(5), (2012).

22. Ods S,Jos.V , "*Azospirillum,* A Free-Living Nitrogen-Fixing Bacterium Closely Associated With Grasses: Genetic, Biochemical And Ecological Aspects", FEMS Microbiol Rev, 24 (4) 487-506(2000).

23. Rousk,K., Sorensen P, "Nitrogen Transfer from Four Nitrogen-Fixer Associations to Plants and Soils", Ecosystems ,(19) 1491 (2016).

24. Allyson, M, Turlough, M, "Genomes of the Symbiotic Nitrogen-Fixing Bacteria of Legumes, Plant Physiology", 144, 615-622 (2007).

25. Michael P, Charles H, "Discovery of Symbiotic Nitrogen-Fixing Cyanobacteria in Corals", Science, 305(5686) 997-1000 (2004)

26. Ghumare, V, Rana, M, Gavkare, O, Khachi, B, "Biofertilizers increasing soil fertility and crop productivity", *Journal of Industrial Pollution control,* 30(2) 196-201 (2014).

27. Bhardwaj, D, Ansari, M.W, Sahoo R.K, Tuteja N, "Biofertilizers function as key player in sustainable agriculture by improving soil fertility", Plant Tolerance And Crop Productivity, *Microbial cell factories,* 10 13-66 (2014).

28. Berruti A, Lumini E, Balestrini R and Bianciotto V,"Arbuscular Mycorrhizal Fungi as Natural Biofertilizers", Let's Benefit from Past Successes, *Frontiers in Microbiology,* 6 (1559) (2016).

29. Schrey SD, Erkenbrack, E, Fruh E, *et al,*"Production Of Fungal And Bacterial Growth Modulating Secondary Metabolites Is Widespread Among Mycorrhiza-Associated Streptomycetes", *BMC Microbiology,* 12(164) (2012).

30. Kipfer T, Moser B, Egli, S, Wohlgemuth T, Ghazoul J, "Ectomycorrhiza Succession Patterns In Pinus Sylvestris Forests

After Stand-Replacing Fire In The Central Alps", *Oecologia*, (2011).

31. Fransson P, "Elevated CO2 Impacts Ectomycorrhiza-Mediated Forest Soil Carbon Flow: Fungal Biomass Production, Respiration and Exudation", *Fungal Biology,* 5(1) 85-98 (2012).
32. Courty P.E, Buee M, Diedhiou A.G: *et al,* "The Role Of Ectomycorrhizal Communities In Forest Ecosystem Processes: New Perspectives And Emerging Concepts", *Soil Biology and Biochemistry*, 42(5) 679-698 (2010).
33. Hayman D.S, "The Physiology Of Vesicular–Arbuscular Endomycorrhizal Symbiosis", *Canadian Journal of Botany,* 61(3) (1983).
34. S Poonia, "Rhizobium: A Natural Biofertilizer", *International Journal of Engineering and Management Research* (1) 36-38 (2011).
35. M A. Baset Mia, Z H. Shamsuddin, "Rhizobium As A Crop Enhancer And Biofertilizer For Increased Cereal Production", *African Journal of Biotechnology*, 9(37) 6001-6009 (2010).
36. M, Javed, "Role of Rhizobium bacteria in bio-fertilizers production", *LAP, Lambert academic publishing,* (1) (2014).
37. D Javier, JS Montana, M Martinez, "Characterization Of Free Nitrogen Fixing Bacteria Of The Genus Azotobacter In Organic Vegetable-Grown Colombian Soils, *Brazilian Journal of Microbiology* 42(3) 846–858 (2011).
38. http://www.authorstream.com/Presentation/purushotam34-918597-azotobacter-as-biofertilizer/
39. S Mehnaz, "Azospirillum: A Biofertilizer for Every Crop", *Plant Microbes Symbiosis: Applied Facets*, 297-314 (2015).
40. http://agritech.tnau.ac.in/org_farm/orgfarm_biofertilizertechnology.html
41. https://foodtank.com/news/2014/02/cyanobacterial-bio-fertilizer-natures-own-solution-forimproved-soil-fertil/
42. JS Singh, A Kumar, AN Rai, DP Singh, "Cyanobacteria: A Precious Bio-resource in Agriculture", Ecosystem, and Environmental Sustainability, *Frontiers of Microbiology* 7 (2016).
43. U Mishra, S Pabbi, "Cyanobacteria: A potential biofertilizer for rice", *Resonance*, 9 (6) 6–10 (2004).

44. Vaishampayan, RP Sinha, DP Häder, T Dey, AK Gupta, U Bhan, A.L. Rao, "Cyanobacterial Biofertilizers in Rice Agriculture", *Botanical Review* 67(4) 453-516 (2001).

45. Khan M.S, Zaidi A, Wani PA, Agron,"Role Of Phosphate-Solubilizing Microorganisms In Sustainable Agriculture", *Agronomy For Sustainable Development*, (27) 29–43(2007).

46. ZaidiA,"SynergisticInteractionsOfNitrogenFixingMicroorganisms With Phosphate Mobilizing Microorganisms",Aligarh Muslim University, Aligarh (2000).

47. Stevenson FJ,"Cycles Of Soil Carbon, Nitrogen, Phosphorus, Sulphur Micronutrients, Wiley, New York, (2012).

48. Boswell E P, R.T Koide, D.L Shumway, H.D Addy, "Winter Wheat Cover Cropping, VA Mycorrhizal Fungi And Maize Growth And Yield". *Agriculture, Ecosystems and Environment. ((*67) 55–65(2001).

49. Bücking H, Shachar-Hill Y, "Phosphate Uptake, Transport And Transfer By Arbuscular Mycorrhizal Fungus Is Increased By Carbohydrate Availability". New Phytologist. 165 (2005).

50. George E,K. Haussler, S.K. Kothari, X.L Li, H. Marshner, "Contribution of Mycorrhizal Hyphae To Nutrient And Water Uptake Of Plants. *In Mycorrhizas in Ecosystems*, ed., D.J. Read, D.H. Lewis, A.H. Fitter, I.J. Alexander. United Kingdom(2010).

51. Grant C, Bitman S, Montreal M, Plenchette C, Morel C, "Soil And Fertilizer Phosphorus: Effects On Plant Supply And Mycorrhizal Development". *Canadian Journal of Plant Science.*(2005).

52. Benizri E, Baudoin E, Guckert A, "Root Colonization By Inoculated Plant Growth Promoting Rhizobacteria", *Biocontrol Science and Technology*. 11 (2001).

53. Hasarin N, Vidaya K,"The study of shelf life for liquid biofertilizer" rom v Panlada, T., Warapom, P., Neung, T., Paul., W. S. and Nantakon, B. "Growth, survival and field performance of bradyrhizobial liquid inoculants formulations with polymeric additives". ScienceAsia 33. 69-77(2007).

54. Phua C.K.H, Abdul Rahim K, Nazrul A.A.W, "Evaluation Of Gamma Irradiation And Heat Treatment By Autoclaving In The Preparation Of Microorganism-Free Carriers For Biofertilizer Products", Jumal

Sains Nuklear Malaysia, Volume 21 (2009).

55. Phua C.K.H, Abdul Rahim K, Nazrul A.A.W,"Shelf life study on Nuclear Malaysia biofertilizer products". In: Nuclear Malaysia Technical Convensyen, Nuclear Malaysia, Kajang, Selangoregetable waste AUJ.T. 11(4)(2009).

56. P Talbot, G Bélanger, M Pelletier, G Laliberté and Y Arcand,"Development Of A Biofilter Using An Organic Medium For On-Site Wastewater Treatment",*Water Science and Technology*(1996).

57. Chen J H, "The Combined Use Of Chemical And Organic Fertilizers And/Or Biofertilizer For Crop Growth And Soil Fertility". *International Workshop on Sustained Management of the Soil-Rhizosphere System for Efficient Crop Production and Fertilizer Use*, 1–11(2006).

58. Hashemabadi D, Zaredost F, Ziyabari M B, Zarchini M, Kavian B, Solimandarabi M J, Torkashvand A M, Zarchini S, "Influence Of Phosphate Bio-Fertilizer On Quantity And Quality Features Of Marigold", *Australian Journal of Crop Science*, 6(6) 101-1109 (2012).

59. Jastrzębska M, Kostrzewska M K, Treder K, Jastrzębski W P, Murkowski P, "Phosphorus Biofertilizers from Ash and Bones-Agronomic Evaluation of Functional Properties",*Journal of Agricultural Science*, 8(6) 58-70 (2016).

60. Igbokwe P K, Asadu C O, Okpe E C, Okoro S E, "Manufacture of Bio Fertilizer by Composting Sawdust and Other Organic Waste",*International Journal of Novel Research in Physics Chemistry & Mathematics*, 2(3) 6-15 (2015).

61. Nayaka S, Vidyasagar G M, "Development Of Eco-Friendly Bio-Fertilizer Using Feather Compost", *Annals of Plant Sciences*, 238-244 (2013).

62. Chang C H, Yang S S, "Thermo-Tolerant Phosphate-Solubilizing Microbes For Multi-Functional Biofertilizer Preparation",*Bioresource Technology*, 1648–1658 (2009).

63. Miranda S Y, Jayadev A, "Different Capacity Of Nutrient Mobilization Of Microbial Flora From A Chemical Fertilizer Applied And Biofertilizers Applied Rhizosphere",*Journal of Global*

Biosciences, 4(2) 1439-1447 (2015).

64. Rokhzadi A, Asgharzadeh A, Darvish F, Nourmohammadi G, Majidi E,"Influence Of Plant Growth-Promoting Rhizobacteria On Dry Matter Accumulation And Yield Of Chickpea Under Field Condition". Am-Euras.*J.Agric. Environ. Sci*. **3**(2): 253-257 (2008).
65. El-Yazeid A.A, Abou-Aly H.A, Mady M.A, Moussa S.A.M, " Enhancing Growth, Productivityand Quality Of Squash Plants Using Phosphatedissolving Microorganisms (Bio Phosphor) Combined With Boron Foliar Spray", *Res. J. Agric.Biol. Sci*,**3**(4) 274-286(2007).
66. Postgate J,"Microbiology Of The Free-Living nitrogen fixing Bacteria, Excluding Cyanobacteria", Currentperspectives In Nitrogen Fixation. *Elsevier/North-Holland Biomedical, Amsterdam*, 217–228 (1981).
67. Young,P, "Phylogenetic Classification Of Nitrogen fixing organisms", *Biological nitrogen fixation*, 43–86 (1992).
68. Sprent J. I,"Nodulation in Legumes", RoyalBotanicGardens, Kew, UK Sprent JI, James EK(2007) Legume evolution: where do nodules andmycorrhizas fit in? *Plant Physiol*, (**144**) 575–581(2001).
69. Huss-Danell, K. "Actinorhizal Symbioses And Their N2 Fixation", *New Phytol* (**136**)375–405 1997.
70. Meeks JC, Elhai, J. "Regulation Of Cellular Differentiation In Filamentous Cyanobacteria Infree-Living And Plantassociated Symbiotic Growth States". Microbiol Mol Biol Rev, **66** 94–121 (2002).
71. Elmerich C, Zimmer W, Vieille C,"Associative Nitrogen Fixing Bacteria",*Biological nitrogen fixation*. Chapman and HallInc, New York, 212–258. (2007).
72. Baldani JI, Baldani VLD,"History On The Biological Nitrogen Fixation Research In Graminaceous Plants: Special Emphasis On The Brazilian Experience", *AnAcad Bras Cienc,* (**77**) 549–579 (2005).
73. "Physiology, Biochemistry and Molecular Biology Of Nitrogen Fixation", *Biology of the nitrogen cycle*, 109–130 (2007)
74. Hu Y, Fay AW, Lee CC, Ribbe MW, "P-Clustermaturation On

Nitrogenase MoFe Protein. *ProcNat Acad Sci USA*(**104)** 10424–10429 (2007).

75. Arnold W, Rump A, Klipp W, Priefer UB, Pühler, "A Nucleotide Sequence Of A 24,206-Base-Pairdna Fragment Carrying The Entire Nitrogenfixation Gene Cluster Of Klebsiella Pneumoniae. *JMol Biol (***203)** 715–738 (1988).
76. Merrick MJ, Edwards RA, " Nitrogen Control Inbacteria",*Microbiol Rev (***59)** 604–622 (1995).
77. Gothwal R.K, Nigam V.K, Mohan M.K, Sasmal D, Ghosh P,"Screening Of Nitrogen Fixers Fromrhizospheric Bacterial Isolates Associated With important Desert Plants",*Appl. Ecol. Environ.Res***6(2)** 101-109 (2007).
78. Bakulin M.K, Grudtsyna A.S, Pletneva A, "Biological Fixation Of Nitrogen And Growth Ofbacteria Of The Genus Azotobacter In Liquid Mediain The Presence Of Perfluoro Carbons*" Appl.Biochem. Microbiol (***4**) 399-402 (2007).
79. Gupta A.K, "The Complete Technology Bookon Biofertilizers And Organic Farming", NationalInstitute of Industrial Research Press India (2004).
80. Wani S.P,"Inoculation With Associative nitrogen fixing Bacteria: Role In Cereal Grainproduction Improvement", *Indian J.Microbiol*, (**30**) 363-393 (1990)
81. Claudine F, Kristina L, Claudine E, "Nitrogen-Fixing Bacteria Associated With leguminous And Non-Leguminous Plants", Plant Soil (**321**)35–59 (2009).
82. Mutch LA, Young JPW,"Diversity And Specificity Of Rhizobium Leguminosarum Biovar Viciae On Wild And Cultivated Legumes:,*Mol Ecol*, (**13**) 2435– 2444 (2004).
83. Lery LMS, von Krüger WMA, Viana FC, TeixeiraKRS, Bisch PM (2008) a comparative proteomicanalysis of Gluconacetobacter diazotrophicusPAL5 at exponential and stationary phases ofcultures in the presence of high and low levels ofinorganic nitrogen compound. Biochim BiophysActa Limpens E, Bisseling T Signaling insymbiosis. Curr Opin Plant Biol (6)343–350 (2003)
84. Tan XY, Hurek T, Reinhold-Hurek B, "Effect of N-fertilization, plant genotype andenvironmental conditions on *nifH* pools in

rootsof rice", *Environ. Microbiol,* (**5**) 1009- 1015 (2003).

85. Krause A, Ramakumar A, Bartels D, BattistoniF, Bekel T, Boch J, Böhm M, Friedrich F, HurekT, Krause L, Linke B, McHardy AC, Sarkar A,Schneiker S, Syed AA, Thauer R, Vorhölter FJ,Weidner S, Pühler A, Reinhold-Hurek B, KaiserO, Goesmann,"A Complete genome of themutualistic, N(2)-fixing grass endophyte Azoarcussp. strain BH72",*Nat Biotechnol,* (**24**) 1385–1391 (2006).
86. James EK, Olivares FL,"Infection And Colonization Of Sugarcane And Other Graminaceous Plants Byendophytic Diazotrophs", *Crit Rev PlantSci* (**17**) 77–119 (1998).
87. Rothballer M, Eckert B, Schmid M, Fekete A,Scholter M, Lehner A, Pollmann S, Hartmann "AEndophytic Root Colonization Of Gramineous Plantsby Herbaspirillum Frisingense",*FEMS MicrobiolEcol* (**66**) 85–95 (2008).
88. Chen WM, Moulin L, Bontemps C, VandammeP, Béna G, Boivin-Masson C,"Legumes Symbiotic Nitrogen Fixation By Beta-Proteobacteria Is wide spread In Nature", *JBacteriol*, (**185**) 7266–7272 (2003).
89. Sprent JI, James EK Legume Evolution: "Where Do Nodules And Mycorrhizas Fit In?", *Plant Physiol*, (**144**) 575– 581 (2007).
90. Lindström K, Martínez-Romero E,"International Committee On Systematics Of Prokaryote ssub committee On The Taxonomy Of Agrobacteriumand Rhizobium: Minutes Of The Meeting", *Int JSyst Evol Microbiol*, (**57**) 1365–1366 (2007).
91. Lindström K, Kokko-Gonzales P, Terefework Z, Räsänen LA "Differentiation Of Nitrogen-Fixinglegume Root Nodule Bacteria", In: *Cooper JE, RaoJR (eds) Molecular techniques for soil andrhizosphere microorganisms*. CABI Publishing,Wallingford, 236–258 (2006).
92. MacLean AM, Finan TM, Sadowsky MJ, "Genomes Of The Symbiotic Nitrogen-Fixing Bacteriaof Legumes, (**144**) 615–622 (2007).
93. Spaink HP,"Root Nodulation And Infection Factors produced By Rhizobial Bacteria", *Annu RevMicrobiol*, (**54**) 257–288 (2000).
94. Yang GP, Debellé F, Savagnac A, Ferro M, SchiltzO, Maillet F, Promé D, Treilhou M, Vialas C,Lindström K, Dénarié J, Promé JC

"Structure Of The Mesorhizobium Huakuii And Rhizobiumgalegae Nod Factors: A Cluster Of Phylogeneticallyrelated Legumes Are Nodulated By Rhizobia Producing Nod Factors With Alpha,Betaunsaturatedn-Acyl Substitutions", *Molmicrobiol*, (**34**) 227–237(1999).

95. Mergaert P, Van Montagu M, Promé JC, HolstersM,"Three Unusual Modifications, A D-Arabinosyl,An N-Methyl, And A Carbamoyl Group, Are Presen ton The Nod Factors Of Azorhizobium Caulinodansstrain ORS571", *Proc Natl Acad SciUSA*, (**90**) 1551–1555 (1993).
96. Gage DJ,"Infection And Invasion Of Roots Bysymbiotic, Nitrogen-Fixing Rhizobia Duringnodulation Of Temperate Legumes", *Microbiol MolBiol Rev*, (**68**) 280–300 (2004).
97. Drlatief Ahmad,"Nitrogen Fixing Biofertilizers; Mechanism and Growth Promotion: A Review", *Journal of Pure and Applied Microbiology* 9(2) 1675-1690 (2008).
98. Abeysingha NS, Weerarathne CS "A Preliminary Study On Quantity Fixation Of Biological Nitrogen Fixation In Sugarcane Grown In Sevanagala In Sri Lanka", J Nat Sci Found Sri Lanka (38) 207–210 (2010).
99. Abreu-Tarazi MF, Navarrete AA, Andreote FD, Almeida CV, Tsai SM, Almeida M "Endophytic Bacteria In Long-Term In Vitro Cultivated "Axenic" Pineapple Microplants Revealed by PCR-DGGE", World J Microbiol Biotechnol (26) 555–560(2010).
100. Akhtar MS, Siddiqui ZA "Role Of Plant Growth Promoting Rhizobacteria In Biocontrol Of Plant Diseases And Sustainable Agriculture. In: Maheshwari Plant Growth And Health Promoting Bacteria", Vol 18, Microbiology Monographs. 157–196(2010).
101. Ali SKZ, Sandhya V, Grover M, Kishore N, Rao LV, Venkateswarlu B "*Pseudomonas* Sp. Strain AKM-P6 Enhances Tolerance Of Sorghum Seedlings To Elevated Temperatures", Bio- Fertil Soils, (46) 45–55(2009).
102. America AHP, Cordewener JHG Comparative LC-MS: A Landscape Of Peaks And Valleys. Proteomics (8) 731–749(2008).
103. Andreote FD, Rocha UN, Araujo WL, Azevedo JL, van Overbeek LS,"Effect Of Bacterial Inoculation, Plant Genotype And

Developmental Stage On Root-Associated And Endophytic Bacterial Communities In Potato (*Solanum tuberosum*)", Antonie van Leeuwen (97) 389–399 (2010).

104. Araujo WL, Marcon J, Maccheroni WJ, van Elsas JD, van Vuurde JWL, Azevedo JL "Diversity Of Endophytic Bacterial Populations And Their Interaction With *Xylella Fastidiosa* In Citrus Plants", Appl Environ Microbiol (68) 4906–4914 (2002).
105. Audenaert K, Pattery T, Comelis P, Hofte M Induction of systemic resistance to *Botrytis cinerea* in tomato by *Pseudomonas aeruginosa* 7NSK2: role of salicylic acid, pyochelin, andpyocyanin. Mol Plant Microbe Interact (15) 1147–1156(2002).
106. Bais HP, Weir TL, Perry LG, Gilroy S, Vivanco JM,"The Role Of Root Exudates In RhizosphereInteractions With Plants And Other Organisms", Annu Rev Plant Biol (57) 233–266 (2006).
107. Baldani VLD, Baldani JI, Dobereiner J,"Inoculation Of Rice Plants With The Endophytic Diazotrophs *Herbaspirillum seropedicae* and *Burkholderia* spp.", Biol Fertil Soils (30) 485–491 (2000).
108. Rossle SC, Urmenyi T, Pereira AR, Silva R, Rondinelli E, von Kruger W, Martins O, Baldani JI, Ferreira PC "Complete Genome Sequence Of The Sugarcane Nitrogen- Fixing Endophyte *Gluconacetobacter Diazotrophicus* Pal5", *BMC Genomics* (10) 450(2009).
109. Bhatia S, Maheshwari DK, Dubey RC, Arora DS, Bajpai VK, Kang SC, "Beneficial Effects Of Fluorescent Pseudomonads On Seed Germination, Growth Promotion, And Suppression Of Charcoal Rot In Groundnut (*Arachis hypogea* L.)",*J Microbiol Biotechnol, (*18) 1578–1583(2008).
110. Bhattacharjee RB, Singh A, Mukhopadhyay SN, "Use Of Nitrogen-Fixing Bacteria As Biofertiliser For Non-Legumes: Prospects And Challenges", *Appl Microbiol Biotechnol* (80) 199–209 (2008).
111. Bilal R, Rasul G, Arshad M, Malik KA,"Attachment, Colonization And Proliferation Of *Azospirillum Brasilense* And *Enterobacter* Spp. On Root Surface Of Grasses",*World J Microbiol Biotechnol* (9) 63–69 (1993).
112. Biswas JC, Ladha JK, Dazzo FB, Yanni YG, Rolfe BG,"Rhizobial Inoculation Influences Seedling Vigor and Yield Of Rice",*Agron J* (92) 880–886 (2000).

113. Bohm M, Hurek T, Reinhold-Hurek B, "Twitching Motility Is Essential For Endophytic Rice Colonization By The N_2Fixing Endophyte*Azoarcus* Sp. Strain BH72",*Mol Plant Microbe Interact* (20) 526–533 (2007).

114. Borneman J, Skroch PW, O'Sullivan KM, Palus JM, Rumjanek NG, Jansen J, Nienhuis J, Triplett EW, "Molecular Microbial Diversity Of An Agricultural Soil In Wisconsin",*Appl Environ Microbiol* (62)1935–1943 (1996).

115. Bothe H, Schmitz O, Yates MG, Newton WE,"Nitrogen Fixation And Hydrogen Metabolism In Cyanobacteri", *Microbiol Mol Boil Rev* (74) 529–551 (2010).

116. Bottini R, Cassan F, Piccoli P,"Gibberellin Production By Bacteria And Its Involvement In Plant Growth Promotion And Yield Increase",*Appl Microbiol Biotechnol* (65) 497–503 (2004).

117. Burbano CS, Reinhold-Hurek B, Hurek T,"LNA Substituted Degenerate Primers Improvedetection Of Nitrogenase Gene Transcription In Environmental Samples",*Environ Microbiol Rep* (2) 251–257 (2010).

118. Burdman S, Jurkevitch E, Okon Y "Recent Advances In The Use Of Plant Growth Promoting Rhizobacteria (PGPR) In Agriculture. In: Rao", Microbial interactions in agriculture and forestry, *Science Publishers,Plymouth*, 29–250(2000).

119. Cartieaux F, Contesto C, Gallou A, Desbrosses G, Kopka J, Taconnat L, Renou JP, "Touraine B Simultaneous Interaction Of *Arabidopsis Thaliana* With *Bradyrhizobium* Sp. Strain ORS278 And *Pseudomonas Syringae*", Pv. Tomato Leads To Complex Transcriptome Changes. *Mol Plant Microbe Interact* (21) 244–259(2008).

120. You Chongbiao Ed, "The Associative Nitrogen Fixation in the Rice Rhizosphere", *Agricultural Publishing House* (1991).

121. FAO, "Legume Inoculants And Their Use", *A Pocket Manual Prepared By Niftal Project and FAO, USA and FAO,* Rome Italy, 63 (1984).

122. Abd Elgadir M, Akanda MJH, Ferdosh F, Mehrnoush A, Karim AA, Noda T, Sarker MZI "Mixed Biopolymer Systems Based On Starch. Molecules", (17)584–597 (2012)

123. Andrade AG, Costa GS, Faria SM Deposição e decomposição da serapilheira em povoamentos de *Mimosa caesalpiniifolia* , *Acacia mangium* e *Acacia holosericea* com quarto anos deidade em planossolo. Rev Bras Ciênc Solo (24) 777–785(2000)

124. Andreote FD, Azevedo JL, Araújo WL, "Assessing The Diversity Of Bacterial Communities Associated With Plants", Braz J Microbiol (40) 417–432(2009)

125. Araújo ASF, Leite LFC, Iwata BF, Lira Junior MA, Xavier GR, Figueiredo MVB "Microbiological Process In Agroforestry Systems.A Review", *Agron Sustain Dev* (32) 215–226(2012).

126. Bahl N, Jauhri S,"Spent Compost As A Carrier For Bacterial Inoculant Production", In: Wuest PJ, Royse DJ, Beelman RB Proceedings of the international symposium on scientific and technological aspects of cultivating edible fungi. The Pennsylvania State University, University Park, 63–68(1986).

127. Bannert A, Kleineidam K, Wissing L, Mueller-Niggemann C, Vogelsang V, Welzl G, Cao Z,Schloter M "Changes In Diversity And Functional Gene Abundances Of Microbial Communities Involved In Nitrogen Fixation, Nitrification And Denitrification In A Tidal Wetland Versus Paddy Soils Cultivated For Different Time Periods. Appl Environ Microbiol (77) 6109–6116, (2011).

128. Bashan Y, "Alginate Beads As Synthetic Inoculant Carriers For The Slow Release Of Bacteria That Affect Plant Growth".*Appl Environ Microbiol*, (51) 1089–1098(1986).

129. Bashan Y, "Inoculants Of Plant Growth-Promoting Bacteria For Use In Agriculture", *BiotechnolAdv* (16) 729–770(1998).

130.

131. Bashan Y, Trejo A, de-Bashan,"LE Development Of Two Culture Media For Mass Cultivation Of *Azospirillum* Spp. And For Production Of Inoculants To Enhance Plant Growth,"*Biol Fertil Soils* (47) 963–969(2011).

132. Beneduzi A, Peres D, Costa PB, Zanettini MHB, Passaglia LMP, "Genetic And Phenotypic Diversity Of Plant-Growth-Promoting Bacilli Isolated From Wheat Fields In Southern Brazil", *ResMicrobiol* (159) 244–250(2008).

133. Bomfeti CA, Florentino LA, Guimarães AP, Cardoso PG, Guerreiro

MC, Moreira FMS “Exopolysaccharides Produced By The Symbiotic Nitrogen-Fixing Bacteria Of Leguminosae”, *RevBras Ciênc Solo,* (35) 657–671 (2011).

134. Bonfante P,“At The Interface Between Mycorrhizal Fungi And Plants: The Structural Organization Of Cell Wall, Plasma Membrane And Cytoskeleton”,*In: Hock B (ed) Mycota, vol IX, Fungalassociations*(2001).
135. BerlinBonfante P, Anca IA,“Plants, Mycorrhizal Fungi, and Bacteria: A Network Of Interactions,”*Annu Rev Microbiol* (63) 363–383(2009).
136. Bonfante P, Genre A (2008) Plants and arbuscular mycorrhizal fungi: an evolutionary- developmental perspective. Trends Plant Sci 13:492–498
137. Borschiver S, Almeida LFM, Roitman T Monitoramento Tecnológico e Mercadológico de Biopolímeros. Polím: Ciênc Tecnol (18) 256–261(2008).
138. Brahmaprakash GP, Sahu PK,“Biofertilizers for sustainability”,*J Indian Inst Sci* (92)37–69(2012).
139. Burity HA, Lyra MCCP, Souza ES, Mergulhão ACES, Silva MLRB Efetividade da inoculação com rizóbio e fungos micorrízicos arbusculares em mudas de sabiá submetidas a diferentes níveis de fósforo. Pesqui Agropecu Bras (35) 801–807(2000).
140. Caetano-Anolles G, Gresshoff M,“Plant Genetic Control Of Nodulation”,*Annu Rev Microbiol* (45) 345–382(1991).
141. Canfi eld D, Glazer AN, Falkowski PD,“The Evolution and Future of Earth’s Nitrogen Cycle”,*Science* (330) 192–196(2010).
142. Chao WL, Alexander M,“Mineral Soils As Carriers For *Rhizobium* Inoculants”,*Appl Environ, Microbiol* (47) 94–97 (1984).
143. Chéneby D, Philippot L, Hartmant A, Henalt, “rDNA Analysis For Characterization Of Idenitrifying Bacteria Isolated From Three Agricultural Soils”, FEMS *Microbiol Ecol* 34:121–128, (2000).
144. Coelho MRR, Marriel IE, Jenkins SN, Lanyon CV, Seldin L, O’Donnell AG “MolecularDetection And Quantification Of Nifh Gene Sequences In The Rhizosphere Of Sorghum (*Sorghumbicolor*) Sown With Two Levels Of Nitrogen Fertilizer”,*Appl Soil Ecol* (42)

48–53(2009).

145. Council NR Biological nitrogen fixation: research challenges. National Academy Press,Washington, DCDakora FD (2003) Defining new roles for plant and rhizobial molecules in sole and mixed plant cultures involving symbiotic legumes. New Phytol 158:39–49 (1994).

146. Deaker R, Roughley RJ, Kennedy IR,"Legume Seed Inoculation Technology – a review,"*Soil Biol Biochem* 36:1275–1288(2004).

147. Smalla K., Sessitsch, A,Hartmann A, "The Rhizosphere: Soil Compartment Influenced By The Root", *FEMS Microbiol. Ecol.,* 56(2): 165 (2006).

148. Kennedy A.C, "Rhizosphere, In: Principles And Applications Of Soil Microbiology", Sylvia, D.M., Fuhrmann, J.J., Hartel, P.G. and Zuberer, D.A. (eds.).,2nd ed. *Pearson, Prentice Hall, New Jersey,*242–262.(2005).

149. Bowen G.D, and Rovira A.D, "The Rhizosphere And Its Management To Improve Plant Growth", *Adv. Agron*., (66) 1–102 (1999).

150. Morgan J.A.W, and Whipps J.M, "Methodological Approaches To The Study Of Rhizosphere Carbon Fl Ow And Microbial Population Dynamics", In: Pinton, A., Varanini, Z. and Nannipieri, P., (eds.)., "*The rhizosphere, Biochemistry and OrganicSubstances at the Soil-Plant Interface"*, New York, USA: Marcel Dekker, 373–409 (2001).

151. Mukerji K.G, Manoharachary C, Singh, J, "Microbial Activity In The Rhizosphere (Soil Biology)", 16 (7), 349.(2006).

152. Kloepper J.W, Zablowicz R.M, Tipping B, Lifshitz R, "Plant Growth Mediated By Bacterial Rhizosphere Colonizers", "*The Rhizosphere and Plant Growth.*" BARC Symp., (14), 315–326.(1991).

153. Be´langer R.R, Avis T.J, "Ecological Processes And Interactions Occurring In Leaf Surface Fungi", In: Lindow, S.E., Hecht-Poinar, E.I., Elliot, V.J. (eds.)., *Phyllosphere Microbiol.,* APS Press, St. Paul, MN, 193–207 (2002).

154. Bayliss C, Bent E, Culham D.E, Mac Lellan S, Clarke A.J, Brown G.L, Wood J.M, "Bacterial Genetic Loci Implicated In The *Pseudomonasputida* GR12-2R3-Canola Mutualism: Identification Of An Exudate-Inducible Sugar Transporter", *Can. J. Microbiol.,* 43, 809–818.(1997).

155. Penrose D.M, Glick B.R, "Levels Of ACC And Related Compounds In Exudate And Extracts Of Canola Seeds Treated With ACC Deaminase Containing Plant Growth-Promoting Bacteria", *Can. J. Microbiol.*, 47, 368–372(2001).

156. Van Loon L.C, "Plant Responses To Plant Growth-Promoting Rhizobacteria", *Eur. J. Plant Pathol.*, 119, 243–254(2007).

157. Tilak K.V.B.R, Ranganayaki N, Pal K.K, De R, Saxena A.K, Nautiyal, C.S. Mittal, S., Tripathi, A.K. and Johri, B.N. "Diversity Of Plant Growth And Soil Health Supporting Bacteria", *Curr. Sci.,* 89, 136–150 (2005).

158. Egamberdiyeva D, "The Effect Of Plant Growth Promoting Bacteria Ongrowth And Nutrient Uptake Of Maize In Two Different Soils", *Appl. Soil Ecol.*, (36) 184–189 (2007).

159. Khan, A.G. "Role Of Soil Microbes In The Rhizospheres Of Plants Growing On Trace Metal Contaminated Soils In Phytoremediation", *J. Trace Elem, Med. Biol.*, 18 355–364 (2005).

160. Babalola O.O, Akindolire, A.M, "Identification Of Native Rhizobacteria Peculiar To Selected Food Crops In Mmabatho Municipality Of South Africa", *Biol.Agric. Hort.*, 27(3–4), 294–309 (2011).

161. Jeyarajan R,Nakkeeran S,"Exploitation Of Microorganisms And Viruses As Biocontrol Agents For Crop Disease Management", In: *Biocontrol Potential and theirExploitation in Sustainable Agriculture,* 95–116 (2000).

162. Chin-A-Woeng, T.F.C, Lugtenberg B.J.J,"Root colonization following seed inoculation", *Plant Surface Microbiology*, 13–33 (2004).

163. Howie W.J, Cook R.J, Weller D.M, "Effects Of Soil Matric Potential And Cell Motility On Wheat Root Colonization By Fluorescent Pseudomonads Suppressive To Take-All", *Phytopathol.*, (77) 286–292(1987).

164. Weller D.M, Thomashow L.S,F.O'Gara D.N, Dowling B, Boesten, "Current Challenges In Introducing Benefi Cial Microorganisms Into The Rhizosphere: Molecular Ecology Of Rhizosphere Microorganisms: Biotechnology And The Release Of GMOs", 1–18(1994).

165. Whipps J.M, "Microbial Interactions And Biocontrol In The Rhizosphere", *J. Experimental Botany.*, (52) 487–511(2001).

166. Hunter-Cevera J.C, "The Value Of Microbial Diversity", *Curr. Opin.Microbiol.*, 1, 278–285 (1998).

167. Abd El-Malek Y, Y. Z, Ishac,"Evaluation Methods Used In Counting Azotobacter", *J. Appl. Bact.*, 331: 269–75 (1968).

168. Adholeya A, A. Prakash," Effect Of Different Organic Manures/ Composts On The Herbage And Essential Oil Yield Of *Cymbopogon Winterianus* And Their Influence On The Native AM Population In A Marginal Alfisol", *Bioresour Technol. Tanu.,* 92: 311–9(2004).

169. Cleyet-Marel J.C, M. Larcher, H. Bertrand, S. Rapior, X. Pinochet, "Plant Growth Enhancement By Rhizobacteria", *In*: Morot- Gaudry, J.F. (ed.), *Nitrogen Assimilation by Plants: Physiological,Biochemical and Molecular Aspects*, pp: 185–99 (2001).

170. Cocking, E.C., "Endophytic Colonisation Of Plant Roots By Nitrogenfixing Bacteria", *Plant Soil*, 252: 169–75(2003),

171. Cohen A.T, P. Mariela, B. Rubén, P. Patricia, "Azospirillum Brasilense and ABA Improve Growth in Arabidopsis",*International, Plant Growth substances Association 19th annual meeting PuertoVallarta*, Mexico-July 21-25, (2007).

172. Dobereiner J, L.E Marrial, M. Nery, "Ecological Distribution Of Spirillum Lipoferum",*Beijerink Canadian J. Microbio.*, (22) 1464–73(1976).

173. Dommelen A.V, V Keijers, J. Vanderleyden, M. De-Zamaroczy, "(Methyl) Ammonium Transport In The Nitrogen-Fixing Bacterium *Azospirillum Brasilense*",*J. Bacteriol.*, 10: 2652–9(1998).

174. Bokhtiar S.M, Sakurai K, "Effects Of Organic Manure And Chemical Fertilizer On Soil Fertility And Productivity Of Plant And Ratoon Crops Of Sugarcane", *Archives Of Agronomy And Soil Science,* 51: 325-334.(2005).

175. Chand S, Anwar M, Patra D.D, " Influence Of Long-Term Application Of Organic And Inorganic Fertilizer To Build Up Soil Fertility And Nutrient Uptake In Mint mustard Cropping Sequence", *Communications in Soil Science and Plant Analysis*, 37: 63-76. (2006).

176. "Supply System On Soil Quality Restoration In A Red And Laterite Soil", *Archives of Agronomy and Soil Science*, 49: 631-637 (2008).

177. Kaur K, Kapoor K.K., Gupta A.P, "Impact Of Organic Manures With And Without Mineral And Soil Science, 168: 117- 122(2005).

178. Kunda B.S, Gaur A.C,"Rice Responses To Inoculation With Nitrogen Fixing And Psolubilizing Micro-Organics", *Plant and Soil,* 79: 227-234(1984).

179. Ratti N, Kumar S, Verma H.N, Gautams S.P, "Improvement In Bioavailability Of Tricalcium Phosphate To Cymbopogon Martini Var. Motia By Rhizobacteria, AMF And Azospirillum Inoculation", *Microbiology Research*, 156: 145-149(2001).

180. Dutta S, Pal R, Chakeraborty A, Chakrabarti K, (2003). "Influence Of Integrated Plant Nutrient Phosphorus And Sugarcane And Sugar Yields.", *Field Crop Research,* 77:43-49 (2002).

181. Tzen Zeng-Yang, Jen-Hshuan Chen, "Evaluation The Application Rate Of Animal Manure Compost In A Strongly Acidic Soil. Fertilizers On Soil Chemical And Biological Properties Under Tropical Conditions", *Journal Plant Nutrition Soil and Environment*, 7: 83- 96(2004).

182. Young C.C, Lai W.A, Shen F.T, Hung M.H, Hung W.S, Arun A.B,

183. "Exploring The Microbial Potentially To Augment Soil Fertility In Taiwan:, In Proceedings of the 6th ESAFS International Conference: Soil Management Technology on Low Productivity and Degraded Soils, Taipei, Taiwan, 25-27(2003).

184. Young C.C, Lai W.A, Shen F.T, Huang W.S, Arun A.B, "Characterization Of Multifunctional Biofertilizer From Taiwan And Biosafety Considerations",International Symposium On Future Development Of Agricultural Biotechnology Park, 373-388 (2004).

185. Mishra D.J , " Role Of Bio- Fertilizer In Organic Agriculture: A Review Research",*Journal of Recent Sciences* Vol. 2, 39-41 (2013).

186. Kyaw M. W, Khin M. S, "Effective Microorganisms In Liquid Carrier Of Biogas Effluent To Improve The Chickpea Yield In Field Trial", *GMSARN International conference on Sustainable Development: Issues and Prospects for the GMS,* 12-1,(2008).

187. Panlada T, Warapom P, Neung T, Paul W. S, Nantakon B, "Growth,

Survival And Field Performance Of Bradyrhizobial Liquid Inoculants Formulations With Polymeric Additives",*ScienceAsia* 33: 69-77 (2007).

188. Phua C. K. H, Abdul Rahim K, Nazrul A. A. W, "Evaluation Of Gamma Irradiation And Heat Treatment By Autoclaving In The Preparation Of Microorganism-Free Carriers For Biofertilizer Products", *Jumal Sains Nuklear Malaysia, 21 (1)*(2009).

189. Phua C. K. H, Abdul Rahim K, Nazrul A. A. W, "Shelf Life Study On Nuclear Malaysia Biofertilizer Products", In: *Nuclear Malaysia Technical Convensyen, Nuclear Malaysia, Kajang, Selangor* (2009),

190. Tran T. N. S, Cao N. D, Truong T. M. G, Tran T. A. T, "Effect Of Co-Inoculants (Bradyrhizobia And Phosphate Solubilising Bacteria) Liquid On Soybean Under Rice Based Cropping System In The Mekong Delta", *Omonrice* 15: 135 - 143 (2007).

191. Thamizh Vendan R, Thangaraju M, "Standardization Of Dosage Of Liquid And Cyst Formulations Of *Azospirillum* For Different Application Methods",*Acta Agronomica Hungarica* 55(4): 475-484 (2007).

192. Vessey J. K, "Plant Growth Promoting Rhizobacteria As Biofertilizers", *Plant Soil* 255 :571 -586(2003).

193. Santos VB, Araujo SF, Leite LF, Nunes LA, Melo JW,"Soil Microbial Biomass And Organic Matter Fractions During Transition From Conventional To Organic Farming Systems",Geoderma, 170; 227–231 (2012).

194. Youssef MMA, Eissa MFM, "Biofertilizers And Their Role In Management Of Plant Parasitic Nematodes. A Review", *E3 J Biotechnol. Pharm Res*. 5:1–6(2014).

195. Raja N, "Biopesticides And Biofertilizers: Ecofriendly Sources For Sustainable Agriculture. J Biofertil Biopestici:, 2013.

196. Araujo ASF, Santos VB, Monteiro RTR, "Responses Of Soil Microbial Biomass And Activity For Practices Of Organic And Conventional Farming Systems In Piauistate, Brazil", *Eur J Soil Biol.*44:225–230(2008).

197. Megali L, Glauser G, Rasmann S. "Fertilization With Beneficial Microorganisms Decreases Tomato Defenses Against Insect Pests.

Agron Sustain Dev:, 2013.

198. Sahoo RK, Ansari MW, Pradhan M, Dangar TK, Mohanty S, Tuteja N, " Phenotypic And Molecular Characterization Of Efficient Native Azospirillum Strains From Rice Fields For Crop Improvement. Protoplasma.(2014).
199. Sinha RK, Valani D, Chauhan K, Agarwal S,"Embarking On A Second Green Revolution For Sustainable Agriculture By Vermiculture Biotechnology Using Earthworms: Reviving The Dreams of Sir Charles Darwin", *Int J Agric Health Saf.*1:50–64(2014).
200. Singh JS, Pandey VC, Singh DP, "Efficient Soil Microorganisms: A New Dimension For Sustainable Agriculture Andenvironmental Development", *Agric Ecosyst Environ,*(2011).
201. Bervejillo J, M.S Dhatt, W.D Gubler,"Evaluation Of Fungicides For Control Of Grape Powdery Mildew",*F&N Tests,* **55**: 94(1999).
202. Bolwell, G.P, & P. Wojtaszek, "Mechanisms For The Generation Of Reactive Oxygen Species In Plantdefence – A Broad Perspective", *Physiological and Molecular Plant Pathology* **51**: 347–366 (1997).
203. Bostock, R.M,"Signal Crosstalk And Induced Resistance: Straddling The Line Between Cost And Benefit", *Annual Review ofPhytopathology* **43**: 545–80 (2005).
204. Colby, S.R,"Calculating Synergistic And Antagonistic Responses Of Herbicides Combinations", *Weeds* **15**: 20–22(1967).
205. Daayf, F, A. Schmitt, R.R. Bélanger, "Evidence Of Phytoalexinin Cucumber Leaves Infected With Powdery Mildew Following Treatment With Leaf Extracts Of *Reynoutriasachalinensis. PlantPhysiology* **113**: 719–727.(1997).
206. Daayf, F, M. Ongena, R. Boulanger, I.E Hadrami, R.R. Bélanger, "Induction Of Phenolic Compounds In Two Cultivars Of Cucumber By Treatment Of Healthy And Powdery Mildew-Infected Plants With Extracts Of*Reynoutriasachalinensis*", *Journal ofChemical and Ecology* **26**: 1579–1593 (2000).
207. Feys, B.J, J.E. Parker,"Interplay of signaling pathways in plant disease resistance", *Trends in Genetics* **16**: 449–455(2000).
208. Fofana B, D.J. McNally, C.Labbé, R. Boulanger, N. Benhamou, A. Séguin, R.R. Bélanger, "Milsana-Induced Resistance In Powdery

Mildew-Infected Cucumber Plants Correlates With The Induction Of Chalcone Synthase And Chalconeisomerase, *Physiological and Molecular Plant Pathology,***61**: 121–132 (2002).

209. Fofana B, N. Benhamou, D.J. McNally, C. Labbé, A. Séguin, R.R. Bélanger, Suppression of induced resistance in cucumber through disruption of the flavonoid pathway. *Phytopathology***95**: 114–123 (2005).

210. Godard, S, I. Slacanin, O. Viret, & K. Gindro,"Induction Of Defence Mechanisms In Grapevine Leaves By Emodin- And Anthraquinone-Rich Plant Extracts And Their Conferred Resistance To Downy Mildew", *Plant Physiology and Biochemistry* **47**: 827–837.(2009).

211. Hammerschimdt R, J. Kuć,"Liginification As A Mechanism For Induced Systemic Resistance In Cucumber", *Physiological PlantPathology* **20**: 61–71 (1982).

212. McNally D, K.V Wurms, C. Labbé, R.R. Bélanger, "Synthesis Of *C*-Glycosyl Flavonoid Phytoalexins As A Site-Specific Response To Fungal Penetration In Cucumber", *Physiological andMolecular Plant Pathology* **63**: 293–303 (2003).

213. Randoux B, D. Renard, E. Nowak, J. Sanssené, J. Courtois, R. Durand, & P. Reignault. "Inhibition Of *Blumeriagraminis*f.Sp. *Tritici*germination And Partial Enhancement Of Wheat Defenses By Milsana", *Phytopathology,***96**: 1278–1286 (2006).

214. Schnee, O. Viret, K. Gindro,"Role Of Stilbenes In The Resistance Of Grapevine To Powdery Mildew", *Physiological andMolecular Plant Pathology,***72**: 128–133(2008).

215. Schneider, S, W.R. Ullrich,"Differential induction of resistance and enhanced enzyme activities in cucumber and tobacco caused by treatment with various abiotic and biotic inducers",*Physiological and Molecular Plant Pathology,***45**: 291–304 (1994).

216. Su H. C, Morgan B, Campbell H, Huang J, Hernandez M.E, Koivunen, P.G. Marrone,"Efficacy Of New Formulations Of Milsana®, Conventional And Organic Regalia®, In Controlling Cucumber Powdery Mildew (*Sphaerothecafuliginea*)",*Phytopatho logy,***99**:S125 (2009).

217. Van Loon, L.C., M. Rep, C.M.J Pieterse,"Significance Of Inducible Defense-Related Proteins In Infected Plants",*AnnualReview of*

*Phytopathology,***44**: 135–162.(2006).

218. Vance C.P, T.K. Kirk, R.T. Sherwood, "Lignification As A Mechanism Of Disease Resistance", *Annual Review ofPhytopathology,***18**: 259–288.(1980)

219. Věchet, L,J. Martinková, M. Šindelářová, L. Burketová,"Compounds Of Natural Origin Inducing Winter Wheat Resistanceto Powdery Mildew (*Blumeriagraminis*f.Sp. *tritici*),*Plant SoilEnvironment,***51**: 469–475 (2005).

220. Wurms, K, C. Labbé, N. Benhamou, R.R. Bélanger,"Effects Of Milsana And Benzothiadiazole On The Ultrastructure Of Powdery Mildew Haustoria On Cucumber", *Phytopathology,***89**: 728–736(1999).

221. Zavareh, A.H.J, A.S. Tehrani, M. Mohammadi,"Effects Of Leaf Extract Of *Reynoutriasachalinensis*on The Defense Responses Of Host In Cucumber-Powdery Mildew Interaction",*Journal ofScience and Technology of Agriculture and Natural Resources,* **11**: 299–307(2007).

INDEX

R

S

T

V